A Practical Treatise upon Modern Printing Machinery and Letterpress Printing

Frederick J.F. Wilson
Douglas Grey

CAMBRIDGE UNIVERSITY PRESS

Cambridge, New York, Melbourne, Madrid, Cape Town, Singapore,
São Paolo, Delhi, Dubai, Tokyo, Mexico City

Published in the United States of America by Cambridge University Press, New York

www.cambridge.org
Information on this title: www.cambridge.org/9781108021548

This edition first published 1888
This digitally printed version 2010

ISBN 978-1-108-02154-8 Paperback

CAMBRIDGE LIBRARY COLLECTION

Books of enduring scholarly value

Printing and Publishing History

The interface between authors and their readers is a fascinating subject in its own right, revealing a great deal about social attitudes, technological progress, aesthetic values, fashionable interests, political positions, economic constraints, and individual personalities. This part of the Cambridge Library Collection reissues classic studies in the area of printing and publishing history that shed light on developments in typography and book design, printing and binding, the rise and fall of publishing houses and periodicals, and the roles of authors and illustrators. It documents the ebb and flow of the book trade supplying a wide range of customers with products from almanacs to novels, bibles to erotica, and poetry to statistics.

A Practical Treatise upon Modern Printing Machinery and Letterpress Printing

First published in 1888, *A Practical Treatise upon Modern Printing Machinery and Letterpress Printing* by Wilson and Grey remains a important work for those interested in the Victorian mechanisation of printing. They list, with illustrations, all the different machines in use in the printing trade, in England and abroad. They outline the development of printing from the early hand presses, and discuss in detail the strengths and weaknesses of the different machines then in use. Information is provided on manufacturers and specifications of the multitude of machinery available for all stages of the printing and publishing process. The book contains valuable information on the development of colour printing, and covers book and newspaper printing as well as the needs of small jobbing firms. It will be of interest to historians of printing and publishing, printers, engineers and industrial archaeologists.

Cambridge University Press has long been a pioneer in the reissuing of out-of-print titles from its own backlist, producing digital reprints of books that are still sought after by scholars and students but could not be reprinted economically using traditional technology. The Cambridge Library Collection extends this activity to a wider range of books which are still of importance to researchers and professionals, either for the source material they contain, or as landmarks in the history of their academic discipline.

Drawing from the world-renowned collections in the Cambridge University Library, and guided by the advice of experts in each subject area, Cambridge University Press is using state-of-the-art scanning machines in its own Printing House to capture the content of each book selected for inclusion. The files are processed to give a consistently clear, crisp image, and the books finished to the high quality standard for which the Press is recognised around the world. The latest print-on-demand technology ensures that the books will remain available indefinitely, and that orders for single or multiple copies can quickly be supplied.

The Cambridge Library Collection will bring back to life books of enduring scholarly value (including out-of-copyright works originally issued by other publishers) across a wide range of disciplines in the humanities and social sciences and in science and technology.

𝔄 𝔓𝔯𝔞𝔠𝔱𝔦𝔠𝔞𝔩 𝔗𝔯𝔢𝔞𝔱𝔦𝔰𝔢

UPON

MODERN
PRINTING MACHINERY.

A Practical Treatise

UPON

MODERN
PRINTING MACHINERY

AND

LETTERPRESS PRINTING.

BY

FRED. J. F. WILSON

AND

DOUGLAS GREY.

Illustrated with Numerous Engravings.

———•••———

CASSELL & COMPANY, Limited:

LONDON PARIS, NEW YORK & MELBOURNE.

1888.

CONTENTS.

Part I.

CHAPTER I.

HISTORICAL.

CHAPTER II.

VARIOUS MOTIONS AND MECHANICAL CONTRIVANCES, PECULIAR TO OR USED IN PRINTING MACHINES.

CHAPTER III.

ONE-SIDED STOPPING-CYLINDER MACHINES.

A 2

CHAPTER IV.

THE TUMBLER OR MAIN MACHINE.

CHAPTER .V.

THE TWO-COLOUR MACHINE.

CHAPTER VI.

THE TWO-FEEDER SINGLE-CYLINDER MACHINE.

CHAPTER VII.

THE DOUBLE PLATEN.

CHAPTER VIII.

PERFECTING MACHINES.

CHAPTER IX.

THE ANGLO-FRENCH MACHINE.

𝔓art II.

CHAPTER X.

MODERN NEWSPAPER PRINTING MACHINES.

Part III.

CHAPTER XXIII.

CHAPTER XXIV.

COLOUR PRINTING.

CHAPTER XXV.

SMALL JOBBING MACHINES.

CHAPTER XXVI.

THE WAREHOUSE.

CHAPTER XXVII.

ON ERECTING AND DRIVING MACHINES.

CHAPTER XXVIII.

POWER.

A Practical Treatise

UPON

MODERN PRINTING MACHINERY

Part I.

CHAPTER I.

HISTORICAL.

The old Wooden Press—The Stanhope Press—Nicholson's Machine—
Kœnig—The first Cylinder Machine—Brown's Machine—Donkin
and Bacon — Kœnig and Bauer's Machine—The first *Times*
Machine — Applegath and Cowper — Curved Stereo Plates—
Cowper's Inking Scheme—Applegath and the *Times* Machine.

"THERE are two sorts of *Presses* in use," writes Joseph
Moxon in 1683, " *viz.*, the old fashion and the new
fashion ; the old fashion is generally used here in
England ; but I think for no other reason than because
many *Press-men* have scarce Reason enough to distinguish between an excellently improved Invention, and
a makeshift slovenly contrivance, practised in the minority
of this Art."

So earnest, indeed, is this worthy man in the cause
of the " New-fashion'd *Press* " that he proceeds to " give
a full description of it, because it is not well known
here in England ; and if possible I would for public
benefit introduce it."

It seemed, indeed, almost time that something should

B

be done in the way of improving upon the Press then in use. Printing had been discovered and practised for nearly two hundred years before even such improvements as graced the " New-fashion'd Press," of which the author of the " Mechanick Exercises " showed so intelligent an appreciation, were introduced. Amongst the devices which are found on the title-pages of the early printers, it pleased one Badius Ascensius, of Lyons, who practised his art between 1495 and 1535, to adopt the Press itself as the sign-manual of his profession, and we have accordingly a number of wood-cuts of different sizes, which represent a press differing in no way from the illustration which Moxon gives of the Old-fashioned Press of his day, and in no essential particular from those constructed of wood which were to be found in old printing offices about fifty years ago.

The " New-fashion'd Press" here spoken of was the invention of one Wilhelm Jansen Blaew, of Amsterdam, with whom Moxon shares the eulogiums he had previously lavished upon his handiwork, " accounting myself so much obliged to his Ingenuity for the curiosity of this contrivance, that should I pass by this opportunity without naming him, I should be injurious to his Memory."

As we have no wish to be injurious to any one's Memory in the matter, let it be said at once that Blaew was originally by trade a joiner, who, having apparently a feeling for his business, was employed by Tycho Brahe, the celebrated astronomer, " for the making his mathematical instruments to observe withal." The developments of Brahe's researches brought him, and Blaew with him, into constant contact with the printers, and the mechanical genius of Blaew, seeing the exceeding clumsiness of the existing press, set to work to " contrive a remedy to every inconvenience." Carried away by the fascination of his new employment, he devoted himself entirely to printing, and having, according to the admiring chronicler, "fabricated nine of these New-fashion'd *Presses*, set them all on a row in his *Printing House*, and call'd each *Press* by the name of one of the Muses."

It is not very easy to see exactly what were the improvements which this excellent man actually introduced, or the inconveniences for which he contrived a remedy. Moxon gives, according to his promise, a detailed account of every part of the New Press, but refers us for a description of the old to the "*Joyners* and *Smiths* that work for *Printers*"—who, unfortunately, are no longer available—and to the cut prefixed to his chapter. The comparison of this with the cut of the Blaew press seems to show that one improvement, at least, was the providing of a handle with suitable running gear for bringing the forme under the platen (it having been previously pushed into place by hand), and a different arrangement of the guides, &c., of the platen head. Curiously enough, a cut of 1819, which gives the wooden press as still existing at that date, though partially, and indeed almost entirely, superseded by the iron presses of Stanhope and others, seems more nearly allied to the *pattern* of the "Old-fashion d," than the "New-fashion'd Press" of Moxon's plates, albeit Blaew's handle is there, of course, and the improvements suggested by him are mostly retained.

THE OLD WOODEN PRESS.

As this wooden press, modified during nearly four hundred years only by the improvements just spoken of is the point from which the discoveries of the present

B 2

century may be said to have started, it will be worth
while to give an illustration and description of it, in
order that the successive improvements may be understood.   The following account is taken from Rees'
" Encyclopædia " (1819), and the press itself, as may be seen
by comparing it with the cut of Badius' press already

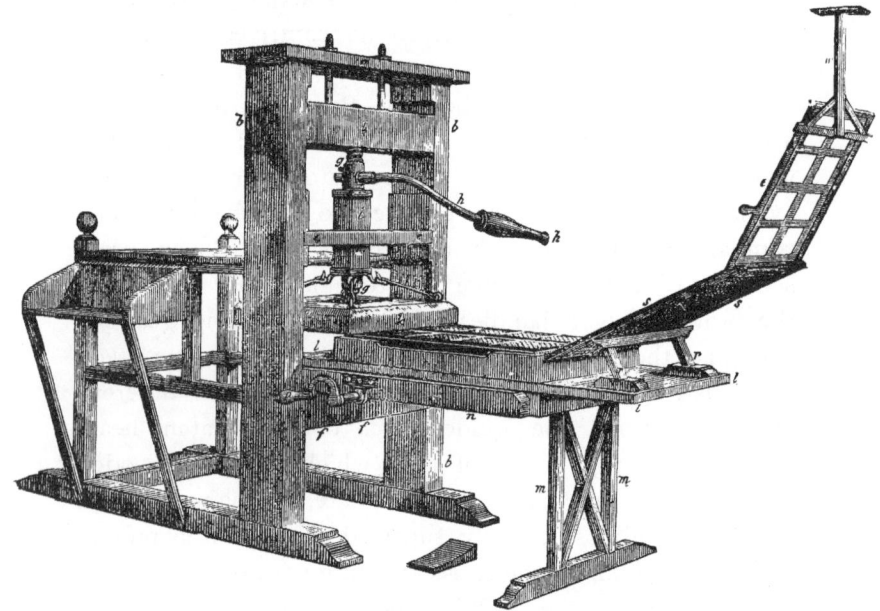

Fig. 1.—The old Wooden Press.

mentioned, is in all essential principles the same as that
used by Gutenberg and Fust in 1455.

The body of the press consists of two strong posts,
b b, placed perpendicular, called the cheeks, which are
joined together by four horizontal cross-pieces.   The
uppermost of these, *a,* is called the cap, and has no office
but to retain the two cheeks at their required distances

asunder. The next cross-piece, *b*, is called the head. It is fitted by tenons at the ends into mortises between the cheeks. These mortises are filled up with pieces of pasteboard, or of wood, in such a manner as to admit of a small motion or yielding. The head is sustained by two long screw bolts, which suspend it from the cap. In the head is fixed a brass nut, containing a female screw or worm, which is fastened in the wood by two short bolts to keep it up. The worm is adapted to receive the screw by which the pressure is produced. The third cross-piece, *e e*, called the shelves or till, is to guide and keep steady a part, *i*, called the hose, in which the spindle of the screw (to be spoken of hereafter) is enclosed. The fourth cross-plank, *f f* called the winter, is fitted between the cheeks to bear the carriage. It sustains the effort of the press beneath, as the head does above, each giving way a little, the one upwards, the other downwards, to make the pull the easier. The spindle, *g g*, is an upright piece of iron, pointed at the lower end with steel, having a male screw formed on its upper end, which enters about four inches into the female screw or worm, fixed in the head. Through the eye of this spindle is fixed the bar or handle, *h h*, by which the pressman works the press. The platen, *k*, or surface which acts upon the paper to produce the impression, is suspended from the point of the spindle by means of a square block or frame of wood, *i*, called the hose, which is guided by passing through the shelves, *e e*. The lower part of the spindle passes through the hose, and its point rests upon the platen, *k*, being received into the plug fixed in a brass pan supplied with oil, which pan is fixed to an iron plate let into the top of the platen, *k*. The pressman then, by pulling the bar, *h h*, fixed in the eye of the spindle *g*, by an iron key, turns the spindle, and by means of its screw presses down the platen upon the forme of type, which is covered with the paper, tympan, and its blankets, all these parts being brought under the platen by the carriage when the impression is to be given. That the platen may be suspended from the spindle, and rise up again with it, the hose, *i*, is attached to the spindle by the garter. This is a fillet of iron screwed to the hose, and entering into a nick or groove formed round the upper part of the spindle, it prevents the hose falling down on the spindle. At each corner of the lower part of the hose there is an iron hook secured, and from these to similar hooks, fastened at each corner of the platen *k*, cords or packthread are looped to suspend the platen, and they are exactly adjusted to hang the platen truly level.

The carriage, *l l*, which is the other principal part of the press, is adapted to run into the space between the cheeks under the platen. It is supported under the ribs, *n*, which are part of a horizontal wooden frame, having its forepart supported by a wooden prop, *m m*, called the forestay, while the other end rests on the winter. On the rails of this frame two long iron bars or ribs are nailed, and under the plank of the carriage are secured short pieces of iron or steel called cramp-irons, which slide upon the ribs, when the carriage is run in or out, by the following means. Beneath the carriage is placed a small spindle called the spit, with a double wheel formed in the middle of it, round which leather girths are passed or fastened, the opposite ends being nailed to each end of the plank, *l*, of the carriage. On the extreme end of the spit is fixed the handle or rounce, *p*, by which the pressman turns the spit, and this, by the means of the wheels and straps, runs the carriage in or out at pleasure. The carriage itself consists of a strong wooden plank, *l*, upon which a square frame of wood is fixed to form the coffin or cell, in which a marble or polished stone is enclosed for the forme of type to be laid upon. To this coffin are fastened leather stay-girths, one to each side, which, being at the opposite ends fastened to the cheeks of the press, prevent the carriage running too far out when drawn from under the platen.

On the forepart of the plank is a gallows, *r r*, which serves to sustain the tympans when turned up from off their forme on the hinges. The tympans, *s s*, are square frames covered with parchment. The frames are made of three slips of very thin wood, and at the top is a slip of iron, still thinner, called a head-band. The two tympans are fitted together by the frame of one being small enough to lie within the other. The outward tympan is fastened with iron hinges to the coffin. Between the two parchments of the tympans two or three thicknesses of blankets are placed, which serve to make the impression of the platen upon the surface of the letters more equal, as also to prevent the letters from being broken by the force of the press. The use of the inner tympan is to confine these blankets. The frisket, *t*, is a square frame of iron, made very thin, also covered with paper or parchment, and fastened to the head-band of the outer tympan by hinges. It folds down upon the tympans to enclose the sheet of paper between them ; the parchment or paper with which the frisket is covered being cut out in the necessary places, that the sheet, when placed between the tympan and frisket, and both together folded down on the forme, may receive the ink from the types in the pages, while the frisket-sheet keeps the margins clean. The tympan and frisket when folded down lie flat upon the forme, and the carriage with them is run into the press; but when the sheet is to be taken out the tympan is lifted up upon its hinges, and rests, as represented, in an inclined position against the gallows, *r r*, before mentioned, at the back part of the carriage; then the frisket, *t*, is lifted upon its hinges and sustained by a slip of wood, *w*, hanging from the ceiling whilst it continues open to take out the printed sheets and put in others.

THE STANHOPE SYSTEM.

The labour of running such a press, it will be easily understood, was very great. With all the modern appliances in the way of rollers, and the advantage of the greater steadiness of the stone, a lithographic hand-press is but a clumsy contrivance. What must have been the labour when, to all the other intricacies of tympan and frisket and the rest, was added the necessity of applying the ink with balls of skin and wool! These balls were made in the form of little wooden cups with handles, the hollows of which were filled with wool or hair covered with a piece of skin nailed to the wood. With them the ink was taken off the ink board or "block," the printer holding one in each hand and distributing the ink by dabbing one against the other; and finally they were applied to the type, a little at a time, until the whole forme was inked satisfactorily—or otherwise. While this process, technically called "beating," was going on, a second man placed the sheet carefully upon the tympan, turning down the frisket upon it, and registering the reverse impression by two pins upon the tympan, which make two corresponding holes in the sheet as it closes. By this time the types are inked; the pressman folds down frisket, tympan, and sheet together upon the stone, and runs it through the press, while his comrade distributes the ink on the balls ; and so on, one removing and replacing the sheets while the other beats, the latter re-inking the balls while the former runs the impression.

Apart from other considerations, the wooden press was too small to print a whole sheet ; and the Apollo press, a French invention, whereby a longer lever was used to bring down a larger platen with this object, was found too fatiguing and heavy in the inking ; and in consequence " they were soon disused, even for printing newspapers, *where expedition is a superior consideration to every other !* " Fancy the expedition possible to a process such as we have described !

We have given a brief description of the wooden press simply because to a history of the successive improvements in printing machinery some understanding of the basis upon which they were built seemed necessary. For the same reason we shall not need to do more than notice shortly a series of important changes in the hand-press itself, which, while entirely changing its character and capabilities of working, did not mark any step in the direction of printing machinery properly so called. Such an improvement was that of Lord Stanhope, whose new pattern practically changed the old wooden press at a bound to the iron proof-press which we now use. His improvements, all-important as they were, left the *principle* of the press where it was. The chief invention which may be claimed for him consists in the application of the so-called Stanhope levers. The press itself was strengthened by the upright frame (called the staple), built in the form of an arch, or sometimes of a vase. In the top of this arch is set the *nut*, which is in fact a box with a female screw, taking

the place of the head of the old press ; while below it,
working either in the neck of the vase-shaped staple
or between uprights, is a piston and cup in which the
toe of the spindle works. To one side of the staple is
a vertical pillar or *arbor*, the top of which is connected
with the spindle by means of the levers already spoken
of, while the insertion of the handle in the arbor, instead
of in the spindle itself as formerly, gives at once the
increase of strength gained by the combination of
levers, and the advantage of a better position and purchase. This, and the return of the platen by the aid
of a balance weight, enabling a heavy iron platen to be
raised and depressed as easily as the old wooden one,
with some minor details—such as the improved slides
for the table, which travels in the ribs instead of over
them, and the drum and girths by which the backward
and forward motion of the table is given—constitute the
main improvements of this press, and, as is to be seen,
leave the machine a hand-press still, with no step taken
towards the application of other power.

The same remark is even more true of the various
ingenious inventions which followed upon the Stanhope
press, and some of which are in use to-day. Before
Lord Stanhope's time, or at about the same period, the
principle of the screw was discarded by Roworth, a
London printer, who applied to the press which he built
the principle of the inclined plane. The spindle carried
a cross-arm at the top, which as it turned was forced
down by acting against the plane fixed under the head.
The screw once brought into question, its various inconveniences were soon acknowledged, and though still used
by Stanhope it has no place in presses after him.
These nearly all adopt the Stanhope coupling system,
and depend for their novelty upon improvements in the

finish and adaptability of various details. The Columbian which was soon abandoned on account of its weight and slowness, gained power by the use of an extremely massive lever taking the place of the screw, and crossing the press from side to side ; the platen being driven down by a heavy square bar of iron attached to the lever in the middle. The Stanhope coupling served to attach a handle to one end of this lever, which was brought back again by a balance weight, on which the American eagle perched triumphant. Hopkinson and Cope followed with the Albion press—so called, presumably, in vindication of the national honour—in which the only notable point is the suggestion of the knuckle-joint. In the Albion press the platen is brought down by forcing a bar of steel to become vertical, the lower end of the bar sliding over the platen head, and the greatest power being obtained at the moment it reaches the perpendicular. The same or a similar method was applied in Cope and Sherwin's Imperial press, which followed, but the knuckle-joint itself first found its way into Harrild's press.

NICHOLSON'S PATENT, 1790.

We come now to the history of a remarkable man, to whose suggestions, if not to his actual achievements, is due, without doubt, the birth of the modern printing machine. The article on "Printing" in the earlier editions of the "British Encyclopædia" was compiled by one William Nicholson, author, inventor, and patent agent, who just a century ago sat in the editor's chair of *Nicholson's Journal of Science*—a publication of high standing in its day—besides keeping a school of science in Soho Square, and occupying his leisure time with such trifles as the prospectus of the British

Institution, which came from his pen. A man of great originality and considerable scientific attainments, he aimed at accomplishing too much, and, to that extent only, failed. The British Institution is held in honour to-day, but people have forgotten who wrote the prospectus. Printing machines throw off hundreds of thousands of copies of the daily papers day after day, but few have ever heard of that famous specification of William Nicholson's, which, issued on the 29th of April, 1790, suggested nearly if not quite *all* the principles involved in them, absolutely *de novo*, and which embraced, as has been well said, more original ideas than any other single patent applicable to printing that was ever granted. Watt has the credit due to the discoverer of the steam engine, though - it took a Stephenson to give his ideas a practical value. Let Nicholson go down to posterity as the inventor of the printing machine, even if it took twenty years more and an "ingenious foreigner" to apply his suggestions in a practicable manner. The specification, to any one who really wishes to understand the principles involved in the invention of the printing machine, is so important, that it shall be given here with very slight omissions.

" My invention," says our author-inventor, " consists in three parts or particulars "—the method of making the types, the application of the colouring matter, and lastly the mode of obtaining the impression.

" I make my moulds, punches, and matrices for casting of letters, in the same manner and of the same materials as other letter founders do, excepting that instead of leaving a space in the mould for the stem of one letter only, I leave space for two, three, or more letters to be cast at one pouring of the metal ; and at the lower extremity of each of those spaces (which communicate by a common groove at the top) I place a matrix or piece of copper, with the letter punched upon its face in the usual way. And, moreover, I bring the stem of my letters to a due form and finish by scraping it in a finishing stick, whose hollowed part is less deep at the inner than the outer side : I call that side of the groove which is nearest the face of the disposed letter the outer side ; and the purpose accomplished by this method of scraping is, that of rendering the tail of the letter gradually smaller the more remote it is, or farther from the face ; and I specify and affirm, that the above described methods of casting two or more letters at once, and of chamfering or sloping their tails, are part of my new inventions.

" I impose or dispose my letter for printing in the common manner, to be used in conjunction with my newly invented improvements ; and I likewise impose it in frames or chases, adapted to the surface of a cylinder of wood or metal.

" In the second place, I distribute or apply the ink or colouring matter upon the surface of the types or originals, aforesaid, by causing the surface of the cylinder, smeared or wetted by the colouring matter, to roll over, or successively apply itself to the surfaces of the said types or originals, or else I cause the said types or originals to successively apply themselves to the said cylinder. I call the said smeared or wetted cylinder, the colouring cylinder ; its surface is covered with leather, or the dressed skins which printers call pelts ; or else it is covered with woollen, or linen, or cotton cloth when the colour to be used is thin, as in calico printing ; and in almost every case the covering is supported by a firm elastic stuffing, consisting of hair, or wool, or woollen

cloth, wrapped one or more folds round the cylinder;
when the covering consists of woollen cloths, the stuffing must be defended by leather or oilskin, to prevent
its imbibing too much colour, and by that means losing
its elasticity. It is absolutely necessary that the colouring matter be evenly distributed over the surface of the
cylinder; for this purpose, when the colour is thick and
stiff, as in letterpress printing, I apply two, three, or
more small cylinders, called distributing rollers, longitudinally against the colouring cylinder, so that they may
be turned by the motion of the latter; and the effect of
this application is, that every lump or mass of colour
which may be redundant, or irregularly placed upon the
face of the colouring cylinder, will be pressed, spread,
and partly taken up and carried by the small rollers
to the other parts of the colouring cylinder, so that
this last will very speedily acquire and preserve an
even face of colour: but if the colouring matter be
thinner, I do not apply more than one or two of
these distributing rollers; and if it be very thin, I
apply an even blunt edge of metal or wood, or other
materials, or a straight brush, or both of these last,
against the colouring cylinder, for the purpose of
rendering its colour uniform. . . . . .

"In the third place, I perform all my impressions
by the action of a cylinder or cylindrical surface—that
is to say, I cause the paper, or cloth, or other material
intended to be printed upon (and previously damped if
necessary), to pass between two cylinders or segments
of cylinders in equal motion, one of which has the
block, form, plate, assemblage of types, or original,
attached to or forming part of its surface, and the other
is faced with cloth, or leather, and serves to press the
paper, cloth, or other material as aforesaid, so as to

take off an impression of the colour previously applied ; or otherwise I cause the block, form, plate, assemblage of types, or original, previously coloured, to pass in close or successive pressure or contact with the paper, or cloth, or other material wrapped round a cylinder with woollen ; or otherwise I cause the last-mentioned cylinder with the paper, or cloth, or other material wrapped round it, to roll along the face of the block, form, plate, assemblage of types, or original, previously coloured ; or otherwise I cause a cylinder having the block, form, plate, assemblage of types, or original, attached to or forming part of its surface, to roll along the surface of the paper, cloth, or other material intended to be printed, and previously spread out upon an even plane covered with cloth or leather ; the said cylinder being supplied with colour by means of a colouring cylinder, hereinbefore described, and hereinafter more particularly to be noticed."

It is hardly too much to say that in these claims is contained everything of real value in the modern machine. Subsequent inventors have done little more than improve upon the details. The substitution of the inking roller for the inking balls, without which the machine was impracticable, would in itself have been an invention of the greatest value ; but so thoroughly had Nicholson foreseen the difficulties which would arise, that he actually added a series of distributing rollers. The cylinder may be said to have been his invention also ; while, in the endeavour to fix his types on the cylinder itself, he was so far ahead of his time as to leap over three generations of inking machines to the application of his principle in the Rotary machines of Hoe, Marinoni, and others.

Unfortunately, Nicholson, though not exactly a

theorist pure and simple, was too little of a practical printer to turn his invention to real account.   His machines, though actually used with some success for wall-paper and pattern printing in colours, was never of any service as a typographic machine in this country. A Dr. Kinsley, however, it is alleged, actually made a machine on Nicholson's principle, which worked fairly well. Of this Dr. Kinsley, however, who was apparently an enthusiastic amateur of printing in Connecticut, we have no further mention, nor any technical descriptions of his machine, which is said to have differed from Nicholson's in being vertical instead of horizontal.*   Savage, in his " Dictionary of Printing," defends Nicholson from the charge of being a visionary, and expresses his belief that had he lived he would have carried his invention to a practical issue.   However that may be, it was reserved for Kœnig, a generation later, to actually set up the first steam press ever worked.

Before this press saw the light, however, one of the principles involved in Nicholson's specification had borne actual fruit.   We have spoken already of the clumsy manner in which the ink was applied to the type under the old system.   The balls were at once the most troublesome and the most necessary articles to the printer, and appear prominently in all representations of ancient printing offices, while they form part of the armorial bearings of the printers' guilds on the Continent. Not only were they difficult to work, but the preparation of the skin, which, with the manufacture of all other

---

* Isaiah Thomas' " History of Printing in America," published in 1810, contains the following reference to Dr. Kinsley (vol. i., p. 129) :— " *Some years since*, Dr. Kinsley, of Connecticut, who possessed great mechanical ingenuity, produced, among other inventions, a model of a cylindrical letter-press." This machine—which was a model of Nicholson's, we are further told—actually was set up and worked.

articles used in the office, fell upon the printers them
selves, was one of the "nastiest processes imaginable,"
to quote a writer who had evidently had experience of the
job, "which converted a press-room into a stinking cloaca."

Besides this, it was clear that if the printing machine,
to which already others besides our ingenious friend of
Soho Square had begun to look, was ever to be invented,
some other manner of inking must be devised. What
this was, Nicholson had already shown. Lord Stanhope,
also, amongst the improvements of which we have
spoken, had attempted to replace the balls by skin rollers;
but the seam, which it was impossible to do away with,
left a mark upon the type at every revolution. With
the hand-press the roller would have been an immense
convenience; to the steam machine it was an indispensable adjunct. The way to the adoption of Nicholson's plan of rollers was paved by the discoveries of
Forster, a practical printer, who, having observed the composition of glue and treacle which the china decorators
used for printing colours upon the biscuit, made use
of a similar compound with which to cover the balls
in his printing office, and the satisfactory result of
the experiment led to the further application of the
process to the casting of a metal cylinder, which was
accomplished by Donkin, an engineer, who produced the
rollers now invariably used in hand-presses, which were
at once adopted by the trade as a perfect substitute for
the old skin balls. In 1810, one year before the erection
of the first steam machine, Messrs. Harrild and Sons commenced the manufacture of composition rollers for the
trade, and established a business of which they had the
monopoly for several years.

The necessities of the times were growing altogether too fast for the capacity of the hand-press. The

C

requirements of newspapers were in excess of the mechanical powers of production. "For daily newspapers in London," says a writer a few years later, describing the machine of Donkin and Bacon, of which we shall speak presently, "expedition is the grand object, as the whole impression (*often as many as* 6,000 *or* 8,000 *copies!*) must be printed between the hours of *midnight and eight or nine o'clock* in the morning; and often when there are important debates in Parliament they are not able to begin printing so soon as midnight."

FRIEDRICH KŒNIG.

To print so enormous a quantity in so short a time (nearly 1,000 an hour if Parliament were late!) obviously required the interposition of a *deus ex machina*, and one was found in Friedrich Kœnig, the son of a small farmer at Eisleben, in Saxon Prussia. He was apprenticed at fifteen to Breitkopf, of Leipzic, when the idea of his machine took shape in his head, and he endeavoured, without success, to obtain encouragement to carry it out. An invitation to take charge of the State Printing Office in St. Petersburg produced no better results as far as the development of his idea was concerned, and in 1806 he determined to visit England, where he at once secured a friend and patron in Thomas Bensley. "We understand," says the writer already quoted, "that a printing machine will very soon be produced by Mr. Bensley, which is the invention of an ingenious foreigner, who has taken several patents for it, but we have not yet had the opportunity of seeing one."

Bensley, in spite of the want of appreciation of the rest of the trade, was so convinced of the practicability of Kœnig's idea, that he found all the necessary capital for experimenting with the proposed machine, and for the next three years Kœnig worked upon it, with the result that in 1810 he took out his first patent for "a method of printing by means of machinery." It seems quite curious that, while Kœnig must have had access to the specifications of Nicholson's patent of twenty years previous, and did in fact eventually adopt all his suggestions with the exception of the rotary type cylinder, which was considered impracticable for many years afterwards, he did not in this first patent apparently see the value of the impression cylinder, one of Nicholson's principal points ; * but contented himself with applying to a platen machine the cylindrical inking rollers which were being already introduced into the trade for use upon the hand-press in successive improvements upon Lord Stanhope's suggestion.

The inking apparatus, which is the chief feature in this 1810 patent, is described in the specification as consisting of several cylinders vertically arranged, above which is an ink-box, through a slit in which the ink is forced by a piston so as to fall upon cylinders, by which it is distributed. The two middle cylinders are for this purpose of different diameters, so "that, when they are revolving, the points of contact may be constantly changed," and for the same purpose an alternating and endwise motion is given to the two cylinders immediately below them, which furnish two inking cylinders revolving in opposite directions. The latter are fitted in a movable frame, and by the action of spiral springs "the one and the other cylinder is alternately applied to the forme." The inking cylinders are

---

* Kœnig did actually read through Nicholson's specification while engaged upon this first machine, but, as will be seen later, apparently discarded the idea as conflicting with his own.

perforated tubes of brass, through the axles of which, also perforated, steam or water is introduced to moisten their felt and leather clothing.

A large portion of the specification was devoted to the "mill-work" which carried the carriage backward and forward, and depressed the plates. This latter operation was accomplished by a compound lever causing a screw to make one quarter of a revolution. The tympan is raised and thrown back, as the carriage leaves the platen, by a chain attached to its end, while a bar depresses it into position again as the carriage returns; the frisket, which has the same centre of motion with the tympan, instead of being hinged to its loose end as in the hand-press, springs up by the action of counter-weights the moment the tympan is thrown back, thus releasing the sheet, which is changed by hand, when the weight of the tympan returning to the forme closes the two together.

The main objection to this machine was the complicated character of its working parts, and the still unsatisfactory character of the inking apparatus. Nevertheless, a machine was actually constructed from this specification, and was set to work in Mr. Bensley's office in April, 1811, upon the *Annual Register*, which was printed at the rate of 800 copies per hour. In the construction of this machine Kœnig had the assistance of Andrew Bauer, an ingenious mechanic and a fellow-countryman. But no sooner had it started working than the inventor commenced fresh experiments with a view to simplification, with the assistance of two London printers, Taylor and Woodfall, and in October of the same year took out his second patent, entitled, "Further improvements on my method of printing by machinery."

The inking apparatus still engaged his attention, and an attempt was made to simplify it by forcing the ink from a cylindrical ink-box, into which a piston was gradually depressed by a screw, through a hole at the bottom, on to two hard rollers which distribute it to others suitably placed. Great stress is laid upon the endwise motion of these cylinders, the irregularities of the ink supply being further counteracted by "a peculiar arrangement" of two rollers placed upon eccentric bearings. The horizontal motion need not be "confined to the middle cylinders. I claim it for the purpose of distributing the ink."

The value of all these arrangements was, however, considerably diminished by the retention of the leather or skin-covered rollers, which are to be found in this and the subsequent machines of 1813 : a curious want of appreciation on the part of a clever inventor, as Mr. R. Harrild had commenced as early as 1810 manufacturing composition rollers for the trade, and printers had made them for themselves some years before.

THE FIRST CYLINDER MACHINE.

The chief interest in this 1811 patent, however, lies in the fact of the first stop-cylinder machine having been constructed from it. The platen has been discarded, probably with a view to simplifying the mill-work (which was, as we have said, far too complicated in the former machine), and in its place is substituted in the centre of the machine "a printing (pressing) cylinder," which is moved in a peculiar manner, making one-third of a revolution for each impression, and then stopping. The cylinder carried three tympans, with iron frames of a peculiar construction to serve as friskets and enclose the sheet, and the surface of the cylinder between these tympans is cut away, so as to allow the forme to pass freely under it on its return. This stopping of the cylinder was necessary to give time for the removal of the sheet, in the absence of any automatic contrivance such as the grippers of modern perfecting machines—the idea of which, by the way, had been suggested, like almost all the modern discoveries, by Nicholson, but not attempted by any of his imitators. The successive motions of the sheet were adapted to this threefold stoppage of the cylinder, the first third of a turn receiving the sheet upon one of the tympans and securing it by the frisket, the second completing the impression and allowing the sheet to be removed, and the third returning the tympan empty to receive another sheet.

The idea of further increasing the speed by feeding at both ends, thus utilising both motions of the coffin, seems to have occurred to Kœnig about this time, for a few lines added to- this specification reserve the right of constructing a double machine on the same principle ; and further a drawing accompanies it, with a design for a circle of machines, the forme moving from one machine to another, which, though impracticable in itself, may have suggested the idea of the multiple machine of Applegath's, which we shall have to describe later.

The cylinder machine was completed in 1812, and was at once put to work, and a number of the trade, including the proprietors of all the morning papers, were invited to Bensley's to see it. Mr. Perry, the proprietor of the *Morning Chronicle*, declared the machine to be "a mere gimcrack," and none of the others appear to have been very much impressed by its capabilities, with the exception of Mr. John Walter, the proprietor of the *Times*, who, after some consideration with the inventor, ordered two of the double machines to be erected in Printing House Square.

These machines were duly erected in 1814; but in the meantime Kœnig had taken out two fresh patents. The first of them, in July, 1813, only deals with the arrangement of the inking cylinders, which had not yet proved satisfactory, and an arrangement of webs and straps for carrying the sheet round the cylinders, thus dispensing with the frisket hitherto employed.

But the succeeding patent of 1814 contains several important points which we may stop to notice. The ductor has taken its proper place at the end of the machine, and is brought in direct contact with the distributing cylinders; "the action of the frame of the lower inking cylinders is simplified by giving a separate frame to each cylinder, with a common centre to both," *i.e.*, placing them transversely across the forme, with their axles meeting on one side. More important still, the rollers are to be covered with composition, and the old skin-covered rollers done away with.

More important still is the new action of the cylinder. It is to have only one tympan, and thus produce one impression only for each revolution—the uncovered parts being of sufficiently small diameter to allow of the repassing of the forme after the impression. Its action, however, is to be *continuous*, and the sheets are to be fed into it at fixed intervals, the stop-action previously given to the cylinder being now transferred to the feeder and tapes connected with it. The continuous motion of the cylinder further suggests the possibility of conveying the sheets to a second cylinder, which shall perform the "reiteration," *i.e.*, print the sheet on the other side, and this is accordingly a feature of the new machine, the sheet being removed from the tympan of the first cylinder automatically by means of an "inner frisket," and by a combination of webs and tapes carried under a second cylinder and impressed with a second impression. Whether a second forme is here employed, or whether the sheet is so turned as to "work and turn" with the one forme, does not clearly appear, but it seems most likely that the latter method was adopted. At any rate, we have here the first perfecting machine, which needed only the improvements of Cowper and Applegath, of which we shall speak shortly, to establish itself firmly in the trade, and demonstrate the entire practicability of printing a perfected sheet by machinery.

BROWN'S MACHINE.

We have passed directly from Nicholson to Kœnig, partly because all previous writers have followed that order, but mainly because the inventions of the latter depend so entirely upon, and must have been to a certain extent prompted by, those of the earlier patentee. But it is worth while to notice—what scarcely any of these writers have mentioned—two patents which John Brown, of whom we have no other record apparently in existence, took out in 1807 and 1809, while Kœnig was still working at his experiments in Bensley's office. The first of these is only interesting from the solution which it more than suggests of the inking problem. The press is similar in construction to the ordinary wooden press, save that, after the impression, the bed carrying the forme "slides out beneath an inking roller . . . . covered with flannel, or any other elastic substance, and then is covered with parchment or vellum, to prevent the ink soaking too far in. . . . . and afterwards is covered with woollen cloth." A large revolving cylinder dips in an ink trough, and after coming in contact with a distributing roller, supplies the inking roller, which revolves and feeds the types, being geared to the spindle which moves the bed. This "inking apparatus (being my principal invention)" may also be applied to ordinary presses by hand, according to the terms of this patent. John Brown must have continued hopeful of success, for shortly afterwards he took out a second patent for improvements in his original design in 1809, just one year before Kœnig completed the experiments he had been working since 1807. This patent describes what must be recognised as a machine, although it was apparently intended to be driven by hand. The forme in this press is fixed, while a carriage runs over it from end to end of the press. When it is over the forme the impression is obtained in one of several ways—either by a screw acting upwards, or by the rising of a rack beneath the forme, or, lastly, by attaching to the underside of the carriage a heavy roller, similar to the galley presses, and thus rolling off the impression. An additional feature, which would seem to have been of distinct benefit in this stage of the press, was the possibility under this arrangement of attaching a tympan and frisket to either end of the forme, and thus feeding alternately at either end of the press, thereby doubling the speed. But the real value must be conceded to lie in the fact that, all these motions of carriage, screw, roller, and the like, being made by the turning of a single rounce or handle, the possibility of applying power became at once evident, and the contrivance is thus elevated to the rank of a machine.

What the fate of this machine was we have not been able to ascertain. Presumably any interest taken

in it was swallowed up in the successive developments
of Kœnig, and Donkin and Bacon, which followed
immediately upon it, and neither of which can very
well have been suggested by it: certainly not Kœnig's,
since we have already shown that he came to England
and commenced work upon his machine before Brown's
patent was published.

DONKIN AND BACON'S PRISMATIC ROLLER.

In the meantime, however, we have lost sight of
Bryan Donkin, whom we last saw engaged with Forster
in the laudable enterprise of transforming the press-room
from a "stinking cloaca" into a chamber redolent of "a
warm scent of ink and paper, anything but unpleasant,"
by means of his composition balls and rollers.   During
the time that Kœnig was experimenting upon Nicholson's
principles, and Brown developing the idea of the platen
machine, Donkin, and a printer named Bacon, were
independently working out the problem of the application
of steam power to the press for themselves ; the result
being that in November, 1813, they obtained a patent
for "certain improvements in the implements or apparatus employed in printing, whether from types, from
blocks, or from plates." This machine apparently
attracted at the time a great deal more notice than
Kœnig's, which, as we have said, was not taken into
favour by the printers. It was publicly exhibited before
the University of Cambridge, and the inventors were
directed to make one for use in printing Bibles and
prayer-books at the University. A contributor to Rees'
"Cyclopædia," writing at about this time, had "examined
the machine at work, and found it to display so much
mechanical ingenuity, and to produce such beautiful

specimens of printing, that we have made a drawing of it."

Kœnig had seized upon the impression cylinder as the keynote of the new printing machine. Donkin and Bacon were caught by the idea of types placed upon the cylinder itself. That these types, with their tapering tails, as designed by Nicholson, would not accomplish the purpose intended, had already become evident. Our inventors endeavoured to overcome the difficulty by imposing the columns of type upon the several faces of a prism, each of which, in turn, as the prism revolves, is brought into contact with the sheet of paper by means of suitably constructed cylinders. The absolute impracticability of this idea is often spoken of by writers on printing, but it is worth while to remark that it in no way differs in principle from the facetted cylinder of Applegath and Cowper's *Times* machine of 1827, which, for some twenty years, worked to the perfect satisfaction of the proprietors of that paper, and which only gave way, in the arrangements of its type columns, to the discovery by Dellagana of the method of curving the stereotype plate now generally adopted on rotary machines. If 'Kœnig's is the first stop-cylinder actually introduced, and Brown's the first platen, Donkin and Bacon may fairly claim to have worked the first rotary machine ; while we have the testimony of an eye-witness that, in contradistinction to either of the other machines, it produced " beautiful specimens of printing." To this no doubt the application of the composition rollers already spoken of must have contributed.

The first idea of the inventors was to bring the prismatic type cylinder into contact with a platen or impression cylinder of a similar, or rather reciprocal pattern, thus insuring an even pressure while the sheet passed under each face in turn ; but experiment showed the difficulty of determining the exact form of this platen, and further trials proved that the ordinary cylinder could be made to adapt itself to the surface of the type by supporting its pivots upon bearings in the manner which had been already used for the inking cylinder, and allowing these bearings to slide in grooves, the cylinder being always kept in close contact with the successive faces of the prism by means of springs or weights acting upon levers.

Besides this rotary motion of the cylinder, it is worth while to notice that Donkin and Bacon in this machine *devised practically the system of inking which is in use to-day.* In place of Kœnig's clumsy upright cylinders, with their perforations and steam moistening, their air-pistons and screws, whose inaccuracy, according to the confession of the patentee, renders " the supply of ink not always quite so regular and uniform as desirable," we have in this machine the ductor and metal roller in no material point differing from our modern appliances. The ink lies between this metal roller and a steel knife, the adjustment of which allows only a small quantity of ink to escape, which is in turn taken off by distributing rollers, and finally by the inking cylinder, all made of composition. If only the number of rollers were greater, and the inking-table adopted, we should have the questions of inking settled once and for all.

The great mechanical skill of Donkin no doubt contributed to bring this machine into greater prominence than it deserved. The excellence of its construction compared so favourably with the clumsy models even of Kœnig, that we are tempted to digress and give him the credit of this much at any rate. He had, it seems, previously made a machine to the order of a certain

Mr. Brightly, of Bungay, in Suffolk, a printer who attempted, with his assistance, to solve the problem then exciting all men's minds, and this, though only one of them was ever made, is alluded to by Hansard,* himself a printer and an expert, as "a beautiful piece of finished mechanism, but the complicated action of the formes, which alternately passed over and under one another in contact with the inking and pressing cylinders, made it what Rowe Moses calls some of his letter-grinders, a *nullibiquarian.*"

Over the Donkin-Bacon machine Hansard is even more eulogistic. "This little machine," he says, "was indeed a most beautiful piece of workmanship." The whole would stand upon an ordinary writing table, and it produced the sheets, at its exhibition, with great rapidity. But to the practical printer it had some grave faults, which Hansard points out, the principal of which was the difficulty of locking up the forme sufficiently tightly to prevent the rotation from dislodging the types—the experiments were all tried with new type fresh from the foundry; and, worse still, as Hansard pointed out to the inventor, one machine would not answer for all kinds of work. "I think I showed him that the various works actually laid on six of my presses in one forenoon would have required four of his machines to execute them"—Bacon having promised that each machine should do the work of eight presses. "This, therefore," he adds, sententiously, "was a speculation not very likely for the London printers to enter into."

Of the future of this machine we do not find any traces. Excellent as were its workmanship and the arrangement of its parts, and promising as it did the element of extreme simplicity, in which it compared

* "Typographia," p. 699.

most favourably with Kœnig's patent, it was destined
to fail, on account, no doubt, of its prismatic arrangement of types. The cylinder was the principle to be
adopted in the press of the future, and everything else
had to make way for it. The superiority of the cylinder
over the hand-press was seen at a glance, inasmuch as
it reduced the *nine* distinct operations which had to be
gone through in handling that machine to the *three*
operations of laying on the sheet, giving the impression,
and taking off the sheet. This much the mere application of the cylinder in place of the platen could do.
The application of steam power reduced the motions to
laying on and taking off, leaving the impression to be
performed automatically by the machinery ; while later
inventions have gradually reduced the laying on to the
mere fixing of a roll of paper at considerable intervals
of time, and the taking off to clearing away a bundle
of sheets delivered automatically by the machine. At
present, however, the cylinder had reduced, at one
stroke, nine operations to three in the case of the hand-
press, while as compared with the prism it was evident
that the circular form, which enables any number of
cylinders to engage with each other accurately, must be
preferable to a shape to which the other parts could
only be accommodated by means of springs and loose
bearings.

To return to Kœnig and his cylinder press. The
first press of 1811, as we have said, was actually set to
work in printing the *Annual Register*, and in April of
that year signature H of that publication was worked on
it to the extent of 3,000 copies. We have no specimens
of this sheet in existence, and no other record of its
success or failure than the significant fact that after the
experiment of these 3,000 copies we hear no more of

the first press.* Something, however, and more than a little, had been accomplished by the experiment. It had been proved that, all the wiseacres to the contrary notwithstanding, it was possible to print by machinery The hand-press had been superseded by the steam-press, and the future history of the printing machine is confined to successive improvements, many of them radical ones, it is true, but none of them going behind that first accepted fact of the possibility of using steam to supply the motive power, of which Kœnig was indeed the prophet and high priest. From that little Saxon town had come the second of two great revolutionary spirits. Luther, born also at Eisleben, had changed the old order of Romanism into the spirit of the new Christianity. Kœnig had already declared that steam was to drive the press, and hand labour had from this act to fall behind in the race. The new order of things began when the first sheet of that long-ago letter H was taken, by Kœnig himself we may believe, from the tympan of that first screw machine.

We may believe that Kœnig and his partners did not quite realise this. To us, who can look at it in the light of modern achievement, the press of 1811 was a triumph; to them it was only a failure, with whatever success lay in it only dimly visible, and so far only partially practicable. We may believe, indeed, that others had seen before this what the partners were now beginning to realise—that nothing satisfactory could come out of any modification of the old screw press.

---

* It must have been this press, however, to which Timperly alludes in the "Dictionary of Typography," where, under the year 1812, is the following memorandum: "1812. The sheets G and Z of Clarkson's 'Life of William Penn' were worked off by an entirely cylindrical press, which, with the aid of two men, worked off eight hundred sheets within the hour."

Already, before the machine was actually put to work, Bensley had invited John Walter, the enterprising proprietor of the *Times*—himself a clever mechanician, and already in great difficulties with the rapidly increasing circulation of his paper—to witness the experiments, and had offered him a partnership in the scheme. But Mr. Walter refused to join them, we may suppose because he believed that no good could come of it; possibly because he looked forward to the cylinder press as suggested by Nicholson; or, indeed, may have heard of Bacon's intention of printing the *Norwich Mercury* by a steam press, and waited to see what came of that.

"I made a point of calling upon Mr. Walter yesterday," writes Bensley to Kœnig in August, 1809, "who, I am sorry to say, declines our proposition altogether, having (as he says), so many engagements as prevent him entering into more."

It may be questioned, however, whether the complete failure of this first machine was not, in fact, a gain rather than a loss to Kœnig. Nothing less would probably have convinced him and his partner that he was on the wrong track, and a success with this *Annual Register* printing would probably have resulted in a waste of many years, if not the whole of the inventor's life, in devising improvements to an already obsolete system. As it was, it had become evident that the screw and platen were an obstacle to obtaining rapid work, which was not at present to be overcome, although the success of the inking apparatus seemed to show that, in this direction, the way had been paved for the application of power.

Fortunately at this moment Kœnig's partners stood firmer, instead of deserting him. Bensley and Taylor were both willing to advance more money, and Bauer was as sanguine of success as ever. The cylinder must at least be tried. A year before the screw machine first saw the light, Kœnig first heard of the Nicholson patent, from the inventor himself, whom he saw in the Queen's Bench Prison, and naturally enough rushed off with Bensley to see the specification. He saw nothing .in it at the time, we may well believe, since he was working on an entirely opposite plan. Indeed, the fact of Nicholson having dared to suggest a cylindrical machine at all, appears to have irritated Bensley as well as himself, for Kœnig himself tells us, in an article written for the "Typographia" of Frankfort, in 1826, that he (Bensley) "could not get through it to the end, declaring it not worth reading," though Kœnig himself read it through, and so "got a notion of what Mr. Nicholson's invention had been."

Kœnig, it must be remembered, always denied that he was in any way indebted to this specification of 1790 for his subsequent idea of the cylinder machine. "*After a few days I had forgotten Mr. Nicholson and his projects.*" This is indeed likely enough, since the patent of the next year, 1811, was, as we have said, upon totally different lines. What does not seem quite so probable is that, this machine having proved a partial if not a complete failure, Kœnig set to work to construct a *cylinder* machine without another thought of the specifications he had carefully read through "only a year ago."

After all, it is a small thing whether in fact the machine of 1813 was inspired by Nicholson or no. It at any rate owed its existence and its life to the skill and ingenuity of Kœnig and Bauer. Nicholson's projects never got much beyond the shelves of the Patent

D

Office.   He himself never appears to have had the
money at his command to bring it out—he died, in fact,
in a debtors' prison ; while Dr. Kinsley, who actually
completed a machine on the principles set out in that
specification, failed to get any work out of it, mainly from
the want of mechanical skill in America to develop it.
Kœnig had both the skill and the money at his command.   He had his inking apparatus practically complete.
He needed only the cylinder, and whether he got this
from Nicholson or no, the next machine was destined
to make a stir in the world.   After a few preliminary
trials there came a day when Bensley and Taylor,
Kœnig and Bauer—Woodfall seems to have left them
by this time—stood by the first cylinder machine in
the office in Frith Street, Soho, anxiously awaiting
the arrival of James Perry and John Walter, the respective proprietors of the *Morning Chronicle* and the
Times, who had promised to inspect the new invention.
Walter had, we know, already declined to embark in
the new enterprise ; there was not much to hope for
from him.   Perry was a better chance, and we may
imagine, if we will, that it was to him Kœnig and
Bauer explained the details of the machinery, while
Walter perhaps stood in the background and talked
to Bensley or Taylor.   But Perry had never believed
in printing machinery, and did not believe in it
now.   He soon took up his hat, and, after a series of
gloomy prophecies and assurances of failure, was gone.
But John Walter, on the contrary, wanted to see
more of it.   This was not the same machine as that
on which he had already pronounced judgment.   We
could imagine our fill over this interview, brimful as
Walter's decision was of consequence not alone to
Kœnig or the *Times*, but to the whole civilised world.

What we *know* is, that when the great man left he had given orders for two machines to be laid down for printing the *Times*.

THE FIRST "*TIMES*" MACHINE.

From this date the history of printing machines becomes in fact the history of the *Times* newspaper, and it is this fact which must be an apology for devoting yet a little space to the story of this first *Times* machine.   A general description of it we have already given ; a more detailed one may be reserved until we come to describe the improved machine of a few years later.   But the story of its introduction into Printing House Square, which is told in the *Times* itself for July 29th, 1847, reads like a veritable chapter of romance.   The machine already erected had been only tried upon book-work.   There were many difficulties to be overcome before it could be set to work upon a newspaper.   Kœnig and Bauer grew disheartened.   The first machine had proved a failure, and now this one was going to fail too.   So utterly hopeless of success did they become, that, after many days of fruitless labour, they one day, as though by a common impulse, "suspended their anxious toil and left the premises in disgust."   Such a confession of failure would have induced most men to throw the whole thing over.   Not so Mr. Walter.   What he actually did we know not, but apparently he worked on at the machine without them ; for we are told that "after the lapse of about three days their retreat was discovered, and they were induced to return, *when they were shown, to their surprise, their difficulties conquered, and the work still in progress.*"

D 2

It is not impossible that Walter was in fact able to give the two Germans some very valuable hints, as the work went on in the building adjoining the *Times* office in which he had installed them. As early as 1804 he had entered with eagerness into the plans of Martyn, a compositor of his own, who had submitted him a working model of what promised to be a power machine. Of the invention itself we have no traces, but Mr. Walter worked steadily at it for some time, and spent a large sum of money upon experimenting with it ; indeed, it is stated that he only gave it up when funds failed him, and his father, who had up till that time assisted him, declined to do so any longer.

Such experience and such energy, combined with the genius and the mechanical skill of Kœnig and Bauer, reaped at last their reward. The machine was finally built, and after many trials it was pronounced ready. In November, 1814, Mr. Walter decided that it would be safe to put its capabilities to a practical test.

There was trouble in the *Times* office too—trouble of a kind of which Mr. Walter had already had experience, and for which he was prepared accordingly. Martyn had suffered great inconvenience and some danger ten years before from the hostility of his fellow-workmen, who had threatened him with the direst vengeance on account of the supposed injury to their craft. His machine had been introduced into the office by stealth, in pieces, and he himself had had to go to work in various disguises to escape the fury of the men. Against such great trouble due precautions had been taken. Kœnig's machine was erected in a building adjoining the *Times* office, with such secrecy that none of the men knew of its existence ; although rumours indeed went the round of the press-room, where the men had openly threatened any one who should endeavour, by his inventions, to interfere with their employment, with "destruction to him and his traps."

"The night on which this curious machine was first brought into use," says the writer of John Walter's obituary in the *Times* of July 29th, 1847, "was one of great anxiety and even alarm. The suspicious pressmen . . . were directed to wait for expected news from the Continent. It was about six o'clock in the morning when Mr. Walter went into the press-room and astonished its occupants by telling them that 'the *Times* was already printed by steam ; that if they attempted violence there was a force ready to suppress it ; but that if they were peaceable, their wages should be continued to every one of them till similar employment could be procured ;' a promise which was no doubt faithfully performed ; and having so said, he distributed several copies amongst them. Thus was this most hazardous enterprise undertaken and successfully carried through, and printing by steam *on an almost gigantic scale given to the world.*" The italics at the end are ours, the actual speed of this first machine being, as appears from Mr. Walter's own leader in the *Times* of the 29th of November, 1814, 1,100 copies per hour—gigantic indeed compared with Kœnig's promises even (he had mentioned 400 an hour in his first contract with Bensley), but equally insignificant by comparison with the editions of to-day.

This editorial of Walter's deserves a place here, and with it we may most fitly close the record of this first working machine.

"THE FIRST NEWSPAPER PRINTED BY STEAM.

"Our journal of this day presents to the public the practical results of the greatest improvement connected with printing since the discovery of the art itself.

"The reader of this paragraph now holds in his hands one of the many thousand impressions of *The Times* newspaper which were taken off last night by a mechanical apparatus.

"A system of machinery almost organic has been devised and arranged, which, while it relieves the human frame from its most laborious efforts in printing, far exceeds all human powers in rapidity and despatch. That the magnitude of the invention may be justly appreciated by its effects, we may inform the public that after the letters are placed by the compositors and inclosed in what is called the 'forme,' little more remains for man to do than to attend upon and watch this unconscious agent in its operations. This machine is then merely supplied with papers, itself places the forme, inks it, adjusts the paper to the newly inked type, stamps the sheet, and gives it forth to the hands of the attendant, at the same time withdrawing the forme for a fresh coat of ink, which itself again distributes to meet the ensuing sheet, now advancing for impression ; and the whole of these complicated acts are performed with such a velocity and simultaneousness of movement that no less than 1,100 are impressed in one hour.

"That the completion of an invention of this kind not the effect of chance, but the result of mechanical combinations methodically arranged in the mind of the artist, should be attended with many obstructions and much delay may be readily admitted. Our share in the event has, indeed, only been the application of the discovery, under an agreement with the patentees, to our own particular business ; yet few can conceive, even with this limited interest, the various disappointments and deep anxiety to which we have for a long course of time been subjected.

"Of the person who made this discovery we have but little to add. Sir Christopher Wren's noblest monument is to be found in the building which he erected ; so is the best tribute of praise which we are capable of offering to the inventor of the printing machine comprised in the description which we have feebly sketched of the powers and utility of the invention. It must suffice to say further that he is a Saxon by birth, that his name is Kœnig, and that the invention has been executed under the direction of his friend and countryman, Bauer."

In this manner was Kœnig's invention ushered into the world : with somewhat of a flourish of trumpets, as, indeed, it deserved. It is a pity that the record of the next few years of his life should be one of quarrels and disappointments. After the success of 1814 he disappears from our history, but we cannot let him go without a few words of farewell. The true facts of his disappointment in England are not difficult to understand. The world had been waiting for the printing machine for years. At the very moment of Kœnig's success with the *Times* Donkin and Bacon had completed their prismatic machine already described. The world accepted Kœnig's contribution gratefully, and to-day hails him as the real inventor of the printing machine, but the world could not afford to stop turning round on that account. Fresh inventors and improvers sprang up on all sides, and, what with protecting his own

discoveries and endeavouring to keep ahead of those of
his rivals, Kœnig must have led a somewhat exciting
existence during the next few years. He was, moreover,
as far as we can discover, of a slow mind, but of a
quick temper, and when finally Bensley, as he thought,
played him false, by encouraging one of his rivals, he
quarrelled with him, and left the country in disgust.

This latter was probably the most sensible thing he
could have done. In Germany he was worshipped as a
hero from another world. They were in no hurry to
invent new newspaper machines, while they were hunger-
ing for the steam press for book-work ; and Kœnig,
retiring to Bavaria, set up a manufactory in the old
convent of Obersell, near Würzburg, in partnership with
the ever-faithful Bauer, establishing. the firm which,
carried on to this day by his two sons, under the title
of Kœnig and Bauer, has grown to be one of the most
important in Europe.

Those who are interested in the private history of
the man, in his early struggles, his failures and success ;
who care to probe to the bottom the " Nicholson
Legend," or decide the merits of the Bensley quarrel ;
who, finally, wish to hear of how he loved in youth, and,
having resigned his lady-love from prudential motives,
married her daughter in later life, may be referred to
Herr Goebel's interesting and appreciative history of his
life, and the excellent series of articles on " The Invention of the Steam Press" in the *Printer's Register* for
1883-4. For the present we must leave him and go
forward.

Kœnig made one last attempt at improving on his
machinery before leaving the country, by arranging to
print both sides of the sheet at once. This first perfecting machine was the natural outcome of the single-

cylinder machine, being in fact a construction only of two of these under one motive power. One large machine of this description was made for Bensley in 1815 ; but it was too clumsy, and, above all, too costly to be of much use to the trade. A great many nice things were said about it, but no orders came in for machines ; and the failure of this attempt was probably the commencement of the misunderstanding between the inventor and his patron which led to the former leaving the country in the same or the next year. In this next year, 1816, first appears on the scene one of the two men who, during the next few years, did so much to make Kœnig's invention practical as a newspaper machine.

APPLEGATH AND COWPER.

The actual patent taken out by Edward Cowper in 1816 did not bear immediate fruit. It depended for its novelty upon the curved stereotype plates which have played so important a part in modern rotary machines, but which were not destined to succeed at first. It is probable that the superiority of printing of the Donkin and Bacon press over the early attempts of Kœnig had suggested a comparison to Cowper and his brother-in-law and business partner, Augustus Applegath, who shared with him the credit of the improvements which during the next few years were patented in the name sometimes of one and sometimes of the other. The advantages of the stereotype plates over the forme of movable types, especially when the imperfections of the printing press added to the difficulty, were very early recognised by printers, and in the sixteenth century Van der Mey, of Leyden, is said to have produced the first stereo block by simply soldering the bottoms of common types together. The expense connected with this method prevented its general adoption, but though it was abandoned after Van der Mey's death the subject occupied the attention of a large number of printers during the next two hundred years. Ged, an Edinburgh printer, took out in 1725 a patent, or privilege, for what seems to have been a development of Van der Mey's method, and he was followed by a series of inventors, mainly French, who during the latter part of the last century met with more or less success in the perfecting of various processes. The idea of the matrix once suggested by Carey, of Paris, it only remained to find a suitable substance, and type metal, clay, and even copper were successively used. It was not, however, until Lord Stanhope, after bringing out his new printing press, turned his attention to stereotyping, that the art was really brought into general use—in England, at all events. This process, to which the name of Andrew Wilson, who worked it out for Lord Stanhope, is usually attached, was productive of some controversy with the University of Cambridge, to whom Wilson offered it, but was eventually adopted both there and at Oxford. The only material difference between the plan then adopted and the present system is that the moulds, instead of being made of papier-maché, were formed of plaster-of-Paris, which, after the types had been oiled to prevent adhesion, was painted over them with a brush, and, after the surfaces had been well covered, poured on to fill a frame in which they were set. From this plaster-of-Paris mould a fresh plate was cast in the way with which we are familiar.

The conviction that Nicholson was right in applying his printing surface to the cylinder itself, must have forced itself upon Cowper. The only difficulty was to arrange the types upon a cylindrical surface. Nicholson's types were utterly impracticable ; they were too trouble-some to cast, and from their construction must have been almost impossible to lock up without causing them to draw out. Bacon and Donkin had tried to get out of the difficulty by flattening the sides of the cylinder into a prism But though their press worked, as we have said, better than Kœnig's early attempts—the proof of which is that the *Norwich Mercury* was being printed by it, while the press at Bensley's lay idle after a brief struggle with the *Annual Register*—it was evident that the peculiar shape of the type prism, and the difficulty of applying rollers and impression cylinders to its shifting surfaces, precluded the possibility of great speed, even if it did not actually bar a more modest achievement. No : if this was indeed the true theory of speed—and that it was has been amply proved by the success of modern rotary machines—it was necessary to revert to the cylinder, and to attack afresh the problem of im-posing the types upon it.

Then, no doubt, occurred to Cowper the idea of the stereotype. Although the plaster-of-Paris mould of course could not be curved, as by Dellagana's method the soft papier-mache is capable of being—it might be possible, though, to curve the plate itself. Accordingly, in January, 1816, we have, according to the specification of Cowper's patent, a new method of printing with curved or bent stereotype plates. " A mould is taken in plaster-of-Paris from the blocks or types used in printing paper, and the plate cast in the usual method practised by stereotype finishers. . . . The plate is then heated equally, ... and laid upon a level board with the face downwards, interposing between the plate and the board some soft substance such as

flannel. The board, the flannel, and the heated plate
thus arranged are passed between two cylinders; a
common rolling press will answer the purpose. . . .
I sometimes curve the plates without heat, but there is
some danger of breaking them. . . The curved
stereotype plates are fixed upon a cylinder, or part of a
cylinder, in the same manner that flat plates are fixed
upon blocks in the common press."

The remainder of the specification describes merely
some improvements in the inking rollers, &c., and the
drawings represent a book-printing machine, in which
the paper is held on to the cylinder by catches, and
preserved from falling by wires bent to the curve of the
cylinder, and at a small distance from it. The suggestion is, however, made that the paper may be held by
endless strings, " as in some machines for ruling paper,"
a suggestion which ultimately developed into the modern
system of tapes.

It seems a little curious that, having once hit upon
the true theory of speed, Cowper should not have
devoted himself to working it out. We have no record
of the achievements of his machine, nor even of its ever
having been constructed, and the next step on his part
was to devote himself to improving Kœnig's system,
without apparently another thought for the type-cylinder,
until in 1831 Applegath took out a new patent for the
same idea. Perhaps the secret of this desertion of his
new discovery lay in the fact that speed, as we are
accustomed to speak of it, was not required at that
time. The six or eight thousand copies of the *Times*,
perhaps increased to twice this number by this time,
did not make any demands upon the mechanician
beyond what was clearly possible to Kœnig's machine;
while the imperfect working of all existing machinery

seemed to suggest that there was work to be done first in the way of getting the best out of the patents already found practicable, without introducing a new system, the necessity for which had not yet been felt.

It was about this time that Kœnig left the country in disgust, possibly accelerated by the suggestions of such men as Cowper, and the readiness shown by Bensley, who saw that he must keep up with the times, to encourage them. When he left, Mr. Walter employed Augustus Applegath, of whom we have already spoken, to superintend the working of the machinery, and under his hands the machine received several important additions, resulting in the increase of the speed from 1,100, at which Kœnig had left it, up to 1,800, and, according to some, to 2,000 copies per hour.

This was something in itself, but a far more important step than the mere perfecting of details was the suggestion of Cowper's, a couple of years later, for an inking table, which at once removed the main difficulty in the way of distributing the colour. This inking table (or "distributing table," as it is called in the specifications of January, 1818) is attached to the forme, and has indentations in its sides which give an endwise motion to "two distributing rollers in a movable carriage, which lies loose on four bearings, and has affixed to it two small friction pulleys." The ink is conveyed to the table by a vibrating roller, which is alternately in contact with the table and with a metal roller (ductor or doctor roller) turning in an ink-trough. The table and forme both pass under the inking rollers, which are three in number, and lie in fixed bearings in the framework, receiving their ink from the table, and inking the forme as it passes under them.

One other improvement also mentioned in this

specification is the method of conveying the sheet from one cylinder to another in a perfecting machine by the introduction of two subsidiary "carrying drums" between the impression cylinders, on which the sheet is carried by means of two sets of endless strings, "each composed of two or more strings kept tight by weights or springs,", the printing cylinders and carrying drums being connected by means of toothed wheels. The perfecting machine erected by Kœnig in Bensley's office, and of which we have already spoken as a failure, was so mainly on account of the impossibility of conveying the sheet accurately from cylinder to cylinder, the register being so imperfect as to render it almost useless. This new feature, in Applegath's and Cowper's hands, however, made a perfecting machine not only possible but easy of construction.

The improvements in the inking apparatus, moreover— and, indeed, the general simplifying of the whole machinery —induced Bensley to apply to the new people for the application of their discoveries to the almost useless machine which Kœnig had left him as the result of all the money spent upon it. "It was not long," says the *Literary Gazette* of October 26th, 1822, "before these gentlemen were requested to apply their inking apparatus to Messrs. Bensley's machine ; and at one stroke, as it were, *forty wheels were removed*, so great was the simplification."

Clumsy and awkward as the old machine was, even with these modifications, it had cost a considerable sum of money, and Bensley continued to use it—probably on that account chiefly—until 1819, when a fire which destroyed his establishment, and did considerable damage to the machine, gave him an excuse for ordering new ones in its place. The old one, however, was repaired, and still lived on for a time, but in 1822 had given place entirely to the new ones, one of them a single-cylinder machine, producing from 1,500 to 1,600 an hour printed on one side, with which Mr. Bensley at that time printed the *Morning Chronicle.* The improvement must have been quite striking, so far as the increased simplicity was concerned, for the writer in the *Literary Gazette* has no hesitation in stating that the original machine contained upwards of one hundred wheels; whereas the new machine, with about ten wheels, accomplished, in point of quantity, exactly the same object, with a marked advantage in regard to the quality of the printing.

So enthusiastic is the writer already mentioned, over a perfecting machine which is capable of throwing off from 800 to 1,000 sheets, printed on both sides, within the hour, that he gravely pronounces that "the printing machine in its present state appears susceptible of little improvement"!! But then we must not forget that Hansard two years later, in recommending his own machine, points out as its chief advantage, that it *supersedes the necessity of steam power.*" There was still something to be learned and done, even in printing machines, in the year of grace 1822.

Applegath, we have said, was employed at this time upon the *Times,* and apparently as each new suggestion occurred to himself or his partner, he applied it as far as possible to the machines already at work there. In this way, during the years which immediately followed Kœnig's departure for Germany, the machine grew out of all recognition; and even before 1827, when the brother-in-law erected an entirely new press, it had reached the form which is known as Applegath and Cowper's machine.

This machine, as improved, for a long time held its own for book-printing purposes. As late as 1854, Messrs. Clowes had twenty-five of them at work, while the *Penny Magazine*, the *Saturday Magazine*, *Chambers's Journal*, and the "Encyclopædia Metropolitana," were all printed by them, to say nothing of the *Magasin Pittoresque*, and *Magasin Universel*, and other foreign publications. The speed acquired by this machine at the time of which we are speaking, was from 2,000 to 2,400 per hour, which satisfied the daily papers. "The machine by which the *Morning Herald* is printed," says a writer in the *Mechanic's Magazine* in 1826, "throws off 2,400 newspapers per hour printed on one side, which is the customary way of printing newspapers, *in order to allow time for the other half of the paper to be composed.*" The saving of time does not seem quite clear, as it must have taken as long to print the second forme as it would have done to print *both* on a perfecting machine; but this plan seems to have satisfied the *St. James's Chronicle*, the *Whitehall Evening Post*, and others, as well as the *Herald*.

All this time, however, the circulation of the *Times* was increasing, and Mr. Walter, in spite of his kindly feeling for Kœnig, which he preserved to the last, found that his present machines, even with all the patching up which Applegath had bestowed upon them, were quite unequal to the work they were expected to do. Bensley had already given in, and one or more of Applegath's and Cowper's latest machines were at work in Frith Street. So thoroughly, indeed, had he identified himself with the new ideas, that the machine we have mentioned is alluded to in the *Literary Gazette* of October 26th, 1832, as "Bensley's machine," a mistake which was perpetuated in the earlier editions of the

"Encyclopædia Britannica" and in Hansard's "Typographia."

So nothing seemed left but for Mr. Walter to write Kœnig a friendly and apologetic letter, and to order an entirely new machine from Applegath and Cowper which should embrace their latest discoveries.

Amongst these it may be convenient to mention here the patent of 1823, in which the system was introduced of placing the distributing rollers, or wavers, diagonally across the forme, thus giving them an end motion.

A large number of other patents were, as we have said, taken out during this time, and previous to the building of the *Times* machine in 1827. Evidently Applegath was not altogether satisfied that they were on the right track, and appears to have tried a number of experiments, all, or nearly all, of which he afterwards discarded: such as the endless flexible inking table—claimed in the patent already alluded to—the method of inking the forme by means of a system of rollers attached to endless bands or chains; and, curiously enough, an apparent attempt to return to Donkin and Bacon's idea by reducing the pressing cylinder to a prism in shape. This last combined the cylinder and platen in a curious way. The sheet was laid upon the topmost side of the prism, and confined by a frisket. It then revolved, and as the sheet reached the lowest point, the coffin and forme were lifted to meet it, and the impression was taken while a fresh sheet was fed on one of the other sides. This presumably was to apply to a hand-press, though it is not so stated, and, so far as we know, no press was ever constructed on the pattern.

Another and an important discovery was that of the rocking cylinder, described in a patent taken out

E

in February, 1824. The idea of this was to bring two cylinders to act alternately, by means of a rocking motion, upon the same forme of types, the latter having to travel through a very short distance before giving the second impression. It was no doubt the success of this experiment that led the way to the principle of multiple impressions adopted in the machine we are about to describe.

The arrangement of this machine, erected for the *Times* in 1827, was somewhat similar in general appearance to the Hoe machine; that is to say, it had four tables, two on each side, the printed sheets being delivered between the two feeding tables on the same side. In order to obtain greater speed, a double set of impression cylinders was used on either side of the machine, composing, in fact, two distinct double machines placed side by side. The forme of type travels the entire length of the machine, being inked at each end, and also in the middle by two extra inking rollers. The rocking system already spoken of is applied to the cylinders, which rise and fall alternately, one of each pair receiving the impression on the outward journey, and the other pair on the return. The proper time for feeding each sheet into the machine was given by the dropping of a roller, called the drop-down roller, which by this action caught the sheet as it was laid to marks on the feeding board. This machine proved a great success. It worked regularly for upwards of twenty years, at a speed of 4,200 an hour; and many years after the new machinery was erected, in 1848, two of this pattern were still employed to print the advertising sheets of the *Times*.

CONTEMPORARY PATENTS.

It must not be supposed that, while Applegath and Cowper were thus bringing out patent after patent, everybody else was idle. Still, though many attempts were made in rivalry to the firm previously to the machine of 1827, that achievement so far threw all others into the shade that it is only necessary to glance at a few of the principal attempts made. Hansard treats most of them with scant courtesy in the "Typographia," possibly because he is reserving himself for a magnificent eulogium, in his best style, of his own press, of which more anon. Thus Mr. Robert Winch, of Shoe Lane, who in 1820 took out a patent for certain improvements on machines or presses, is dismissed with the remark that "upon comparing this apparatus with Kœnig's inventions and that of Rutt"—which was really a hand machine working upon Kœnig's principle—"it would be difficult to discover on what part of the plan the present patentee founded his claim of novelty and originality. This machine was advertised, exhibited, offered in shares, and finally disposed of somehow or somewhere ; but the how or the where I never could find out !"

In this same year (1822), Samuel Cooper and William Millar patented a machine with the imposing title of the "British and Foreign Press ; " but Hansard will have none of them. "Any person having a tolerable knowledge of what printing machines are, or ought to be, who will bestow close attention to the tedious and perplexing description in which all the parts, both old and new, are referred to, without distinction, by a confusion of several hundred large and small letters and

E 2

numerical figures, will soon perceive that this is not the *ne plus ultra* of printing machines."

Mr. Bond, of West Street, Bermondsey, is no better, it seems, and his machine patented in 1823 is described as "equally complicated, and possessing about as great a portion of originality as the British and Foreign machine." There are, by the way, several good points in this specification; but the machine is too complicated, and contains too many parts, which was probably the cause of its non-success.

A TWO-COLOUR PRINTING MACHINE.

Sir William Cowper, too, about this time was, amongst other things, engaged in working on a machine for printing in two colours, the chief feature of which was the motion of the cylinders over the forme, which remained stationary. This machine was, we believe, actually applied to printing stamps, &c., on bankers' promissory notes. Its description and mechanism, however, are interesting, and repay a dip into the *London Journal of Arts and Sciences* for 1823, or the work of Hansard's already quoted.

In 1820, Thomas Parkin had invented an inking apparatus for hand-presses somewhat similar to Applegath and Cowper's, and a little later attempted to bring out a machine in which this was the principal feature, but we hear nothing further of its success, and probably whatever good was in it was absorbed in the latter patents of the *Times* machines shortly to be described.

NAPIER'S MACHINES.

The above patents, it will be seen, were nothing more than successive attempts, more or less successful, to apply the principles laid down by Kœnig and Applegath to some modified form of their machine. The machine which old Hansard introduces with such a flourish of trumpets as his own—though he had previously found fault with Bensley for appropriating the credit of his inventor's genius in a like way—had, however, two special features. This machine was built for Hansard by D. Napier,* better known a little later for his reintroduction of the platen machine after Kœnig had abandoned it. We have already spoken of the rocking cylinder as being one of Applegath's improvements. This machine, however, would appear—though there is some difficulty in getting at the precise date of its building—to have been at work at least as early as any machine of Applegath and Cowper's containing it, and certainly some years before the *Times* machine of 1827, in which we first noticed this feature. In point of fact, these early patentees appear to have found it almost impossible to hold their improvements. Hansard himself remarks of another machine which apparently encroached upon one of Applegath's inventions that " I daresay it will never be contested, for I never yet saw a printing machine which did not, in some part or other, bear a strong similarity to other inventions or patents for the same purpose ; and they have thus become so involved in each other's ideas, that an inquirer has only to investigate the various printing machines from the time of Nicholson, and he may overturn the exclusive right, as a whole, to any subsequent machine." At any rate, this cylinder press of Napier's contained not only the rocking cylinder, but a more important

---

* Napier is credited with having built the first machine ever set up in Ireland, for the *Dublin Evening Post*, though steam was not introduced till 1833, when Mr. Gunn constructed a steam machine for P. D. Hardy, which was followed by the *Evening Mail* machine.

improvement still in the introduction of the "grippers," which, curiously enough, were not adopted by any of the other contemporary machines. A capital description and plate of this machine are to be found in Karl Faulmann's "Illustrirte Geschichte der Buchdruckerkunst," p. 672.

A good deal of this no doubt Kœnig had felt, and it was this eager rivalry which drove him from the country when he found himself unable to keep in step with the rapid march of progress which followed the publication of his great discoveries. The only chance for an inventor was to *keep* ahead, and this Applegath and Cowper alone really succeeded in doing. They had already brought the cylinder machine to a point of perfection of which we should not be ashamed to-day. The rotary machine alone remained untouched, and to this they were next to turn their attention.

Before describing this last step in printing machinery, the success of which brings us within measurable distance of our own time, it may be well to devote a word to Napier's Platen machine. Satisfactory as the cylinder was, it required more careful handling than would suffice under the old system. Given this, the cylinder worked all right; but in its absence—and in those early days of the machine skilful manipulation was difficult enough to procure—the old press made a better show than the new. Steam, however, in spite of Hansard and Johnson, the latter of whom wishes Parliament and the British public to assist him in exterminating the steam machines body and breeches,* was evidently to be the

---

* With a curious qualification : "For the sake of humanity (!) there is no one, we believe, that would object to the adoption of these machines for newspapers of an extensive circulation, provided the proprietors preferred the loss from the destruction of their type,

motive power of the future; and the many who sighed
for the old-fashioned system began to canvass the possibilities of applying steam to a platen machine. For
these, then, Napier designed the machine which to this
day, though slightly modified by modern improvements,
bears his name, and which differed so little from that
with which we are familiar (and which is described in
another chapter) that we will not dwell upon it here.
Suffice it to say that William Spottiswoode, who had
about this time set up a steam engine, with one of
Applegath and Cowper's cylinder machines, introduced
by its side a Napier Platen, worked by the same engine,
which met with Hansard's warm approval, possessing,
in his opinion, "the principle by which better work
may be effected than by any of the others, but not
with such rapidity. The impression is by a platen
moving vertically; it has a self-acting tympan and frisket,
and appeared to me doing at the rate of six or seven
hundred per hour, one side, in excellent style of work."

To return now to the *Times*, and Applegath and
Cowper's last improvement. In 1848, after having given
several years to experiments on the rotary principle of
Nicholson and Bacon, they erected a machine which was
in many respects the most remarkable ever made. It
was successful in its working from the first, and marks
the commencement of a new era in newspaper printing.
The following description is taken from the account in
the "Encyclopædia Britannica." A more detailed account
yet, for which we have not space here, may be found
in Weale's "London and its Vicinity,"* to which we
may refer the curious.

&c., to the advantage derived from them in point of time."—*Johnson's
Typographia*, ii. 660.

* Reprinted in pamphlet form, "A Description of Applegath

" In the centre of the machine is a vertical cylinder or drum, 5 feet 4 inches in diameter. In contact with it, and revolving each on its own vertical axis, are eight impression cylinders, 13 inches in diameter each of which has a set of inking rollers working in advance of it. The cylinders move with the same velocity as the surface of the drum. The columns of type are placed in a kind of iron galley, or *turtle*, curved to fit the surface of the drum. The outer surface of these galleys is not formed into a segment of a circle, but into facets, each the width of a column ; the wedge-shaped interval, which is left between the top and bottom of the types of any two adjoining columns, is compensated by column-rules, made thicker at the top than at the bottom in the same proportion. The middle column-rule is fixed. The columns are locked-up in the galleys by means of screws, and the column-rules press the types together like key-stones in an arch. The fixed rule in the centre prevents the types from rising. The galleys are then screwed on the drum, the columns vertical. The outer face of the formes is now, it must be remembered, a series of facets—sides, as it were, of a polygon ; the surfaces of the impression cylinders are made to conform to these facets with sufficient accuracy by paper overlays. When stereotype plates are used, they are cast by Dellagana's process in accurate segments of a circle, and the overlay is unnecessary. The formes of type do not, of course, occupy the whole circumference of the central drum— a large part of the remainder is made the inking table.

and Cowper's Horizontal Machine for Printing the *Times*," &c. &c., and containing a brief summary of all the improvements in the *Times* machine since Kœnig. See also Bohn's " Pictorial Handbook of London."

The inking table precedes the type formes, and supplies ink to a ductor roller which works between two straight edges. As the drum revolves, a portion of ink is taken from the ductor by two vibrating rollers, and distributed on to the inking table. The inking table precedes the type formes, and, as it passes the inking rollers attached to each impression cylinder, comes into contact with it, and receives ink from its surface. The type formes following next come into contact with these inking rollers, and take from them the ink they have just received. The inking table passes under the impression cylinders without touching them ; but the type is brought into contact with the paper upon them, and the impression is given. Therefore, at every revolution of the drum the type is inked eight times, comes into contact with eight impression cylinders, and prints eight sheets of paper.

" It is most difficult to convey by any verbal description the singularly ingenious mechanism by which the sheets of paper are conveyed to and round the impression cylinders. It must be remembered that the sheets are necessarily laid on the feeding table *horizontally*, and that they pass around the cylinder *vertically*. The task will be rendered somewhat simpler by reminding the reader that each impression cylinder is a complete machine within itself, acting with the drum, but independent of the other cylinders, and that, as each has its own system of inking rollers, so each has its own system of feeding drums and tapes. The white paper is laid on the feeding table at the top ; each sheet is placed by the layer-on to the centre of the feeding drum. At the right moment the sheet is advanced by finger rollers until its forward edge is brought between two small rollers, each connected with a series of endless tapes, between which it is passed vertically downwards. At the right moment

its further progress is arrested by two vertical slips of
wood called 'stoppers,' which start forward and press
the sheet against two fixed stoppers, and at the same
moment the two rollers and their tapes separate, and
leave the sheet extended vertically between the two pairs
of stoppers. Observe that, up to this moment, the travel
of the sheet has been vertically downwards, and that its
plane surface is part of a radius from the axis of the
central drum. The problem now to be solved is to give
it a horizontal movement towards the centre, preserving
its vertical position. The instant the sheet is arrested
vertically between the stoppers its top edge is caught by
two pairs of small finger or suspending rollers ; at the
same time the stoppers separate, and the sheet is suspended for a moment between these rollers. A slight
inward motion is then given to the suspenders, sufficient
to bring the inner edge of the sheet into the mouth of
two sets of horizontal tapes, by which it is carried round
the impression cylinder and printed. As the sheet, after
being printed, issues from the horizontal tapes, it is
delivered to other sets, by which it is conveyed outwards, under the laying-on board. Arrived at the proper
point, it is again caught at the top edge between
suspending rollers, the tapes _separate, and it hangs for
a moment, when the taker-off, who sits below the
layer-on, releases it by a slight jerk, and lays it on his
board.

   " No description can give an adequate idea of the
scene presented by one of these machines in full work
—the maze of wheels and rollers, the intricate lines of
swift-moving tapes, the flight of wheels, and the din
of machinery. The central drum moves at the rate of
6 feet per second, or one revolution in three seconds, consequently 15 sheets are printed in that brief space. The

diameter of an eight-feeder, including the galleries for the layers-on, is 25 feet."

The *Times* before 1848 employed two of these eight-cylinder machines, each of which averaged 12,000 impressions per hour, and one nine-cylinder, which printed 16,000.

Sic transit gloria. This machine, too, is numbered with the dead, and with it our history of the past may appropriately close. With the Hoe machines which replaced it at the *Times* office commences the new era of printing machinery, which must be dealt with in another chapter. In spite of Nicholson's failure to carry out his ideas of the web machine—in spite of Hansard's wish to get rid of the inconveniences of steam, and Johnson's appeal to the public to discountenance the introduction of machinery into the trade ;—nay, though the Encyclopædia man stands aghast at the attempts to describe what is now obsolete, and by implication refuses to believe in any further development ;—despite all these failures and prognostications, the newspaper machine of to-day is an accomplished fact : accomplished, too, upon the lines originally laid down by the old enthusiast of Soho Square just one hundred years ago. The great improvement in the modern machines is found in the economy of space and of labour, for the latter of which the web principle is responsible. In Part II. we shall endeavour to describe all the principal machines in use, taking up the history of their development from the point at which this chapter leaves it. Meanwhile, in order to prepare our readers to thoroughly understand the somewhat technical descriptions which follow, we proceed to the discussion of the mechanical principles which are involved in the construction of Modern Printing Machinery.

CHAPTER II.

VARIOUS MOTIONS AND MECHANICAL CONTRIVANCES, PECULIAR TO OR USED IN PRINTING MACHINES.

The Lever—Wheel-and-Axle—Inclined Plane—The Wedge—The Screw—Pulleys—Gearing—Cranks—The Eccentric—The Cam —Levers—Bed Cranks—Knuckle and Toggle Joints—Racks— Parallel Motion—Universal Joint—Reversal Motion—Rocking Frame—Tumbler—Brake—Balance Weights.

To the intelligent custody of any piece of machinery, there is necessary, not only a practical acquaintance with the working of its different parts, but some general idea of the principles upon which these parts move, and of the various relations between the different pieces of metal which compose the machine, and which serve to transmit or change the direction of the power by which the whole is set in motion.

No one can look at even the simplest machine without noticing for himself something of this change of direction or character ; for example, the change from circular motion in the revolution of the shafting and pulleys, by which all printing machines are primarily set in motion, to the travel, or the backward and forward motion, of the table ; the rise and fall of the platen ; or the motion, often itself irregular, of the cylinder—are all instances of the transmission of power in altered and continually changing forms, which evidently need a few words of general explanation, in order that their nature and origin may be understood.

The principles of mechanics involved in the study of the mechanical *forces* belong too clearly to the realm of pure mathematics to be in place in a treatise like the present. The consideration of the mechanical *powers*, on the contrary, is indispensable to the proper understanding of machinery ; and although it is perhaps most usually treated mathematically, as intimately connected with, and indeed dependent upon, the laws for the resolution of forces, it has another or practical side, which is equally capable of explanation, without more than a passing reference to these laws.

The mechanical motions with which we shall deal may be divided into two classes : those which change the direction or character of the motion, and those which augment or diminish its force, it being remarked that the latter class, by a well-known law of mechanics, diminish or increase the speed in inverse proportion to the gain or loss in power. That is to say, that in a pair of gearing wheels, for example, of unequal sizes, in which the smaller wheel moves three times as fast as the larger, the larger wheel will, by way of compensation for its slower movement, be capable of exerting or transmitting three times the power of the smaller. All mechanical powers so called, except the fixed pulley, belong to the former of these classes.

Many motions—indeed, most—are of a complex character, and, partaking of the nature of both these classes, at the same time alter the direction or character of the motive power, as well as increase or diminish its intensity. The ordinary lever, with its various combinations, usually acts as a multiplying power, as well as a mere means of changing the direction of a force. One at least of these characteristics, however—change of *direction*, *character*, or *intensity* — is impressed upon every motion used in machinery.

The first and last of these, the simple change of

direction, or the increase and decrease of the intensity of a force or power, are sufficiently easy to understand, or at all events may be readily appreciated by studying them as they occur in practice in the machines with which we are most of us familiar. The *character* of any given motions met with, deserves a few words before we go further.

The only practical—as it is the only philosophical— method of studying the character of any motion, is to study the movement of a single point. This point may be considered as moving in one of three ways :—

(1) It may move in a straight line. This movement is of course perfectly constant in direction, being that of the line in which it moves.

(2) It may, while remaining in the same plane—*i.e.,* travelling constantly upon a real or imaginary flat sur- face—yet change the direction of its motion, and travel in a curved line, as of course does every point upon the circumference of a circular wheel.

(3) Lastly, it may move, without reference to a fixed plane, in a curve of any kind. Such a motion is that of many cams, eccentrics, and the like, certain points upon which travel‐ in curves which change their plane of motion constantly.

It will be the purpose of this chapter to discuss many cases of the interchangeability of these three primary cases of motion, and their transference from one part of a machine to another.

It will be seen at once that a certain form of the second case, that of pure circular motion, is the point from which, in the discussion of printing machinery, all our investigations must necessarily commence, since this is the form in which the continuously revolving shafting of the machine-room first presents the motion to us, which is ultimately to be resolved into the various motions of the printing machines which work from it.

In this place it will only be necessary to notice two methods of transferring this motion. Firstly, we may transfer the circular motion of one piece to another without changing its *character*. If, for example, the circumference of a wheel revolving in any given direction be placed in close contact with the circumference of a second wheel in the same plane, the motion will be transmitted from one to the other, so as to cause the second wheel to revolve in the opposite direction. In practice, the two wheels are furnished with teeth to prevent their slipping, but the effect and the principle are alike identical in the two cases, and we have in this pair of wheels a simple instance of the transmission of circular motion from one piece to another. The same effect may be produced by connecting the two wheels by means of a cord or belt, which passes round a portion of the circumference of each. In this case the circular motion of the moving wheel is transferred to the cord, which lies close to its circumference and moves with it, and in turn the now moving cord transmits its motion without change to the second wheel, with which it lies in contact in a similar manner. It is only necessary to note that the wheels move in the same direction where the cord is carried straight round both, and in opposite directions—the same as where their circumferences actually touch—only when the cord is crossed between the wheels. The practical application of these methods will be discussed more fully when we come to speak of pulleys.

With regard to the change of circular motion into motion in a straight line, we may take one single instance before proceeding further. It is evident that a point upon the circumference of a circle may be considered as travelling with reference to any given diameter of the circle, starting from one end of this diameter and continually making towards the other, though by a somewhat circuitous route, reaching it at last, and then returning to its original position by a similar path on the other side. If, then, we take any given point upon the circumference of a moving wheel, and attach it by means of a rod or otherwise to the end of a straight bar compelled by guides to move only in the direction of one of the diameters of the circle ; or, in the alternative, set in this point a pin, which, by means of an arm-piece at the end of the bar, continually pushes it forward or backward, while preserving its own freedom of motion, it will only need a moment's consideration to see that the bar will travel backwards and forwards a length equal to the diameter of the circle or wheel. The actual application of this principle in the crank and various forms of the cam will be seen as we proceed to discuss the mechanical motions in detail.

We have now arrived at a sufficient idea of what species of motions we are likely to have to deal with, to turn our attention to the ordinary mechanical powers which, with their modifications, are found, more or less, in all machinery.

The mechanical powers are usually reckoned as six in number—the *lever*, the *wheel-and-axle*, the *pulley*, the *inclined plane*, the *wedge*, and the *screw*. All of these, with the exception of the wedge and the inclined plane, are familiar to the machinist in their simple forms, and even these two, which are in fact one in origin, form the rudiments of several mechanical motions, such as the cam or shape, of which we shall speak later.

The Lever.—The lever is a rigid bar capable of

turning about a fixed point which may be situated anywhere within its length. This point is the *fulcrum* of
the lever, and the portions of the bar on either side
are called the *arms*. These arms may be either in a
straight line or inclined towards one another ; in the latter
case it is known as a " bent lever." The effect of the lever
is determined by the position of the fulcrum. When
this is placed in the middle the force on either arm is
of course equal, and the lever merely acts by way of
changing the direction. The further, however, the fulcrum
is from the power end (*i.e.*, the end at which the initial
power is applied), the greater will be the resulting force,
a fact familiar enough to every one in the use of the
ordinary crowbar in lifting.

The Wheel-and-Axle.—This is almost identical in its
action with the lever, the only difference being that the
power may be exercised during a considerable time. It
consists of an ordinary shaft with a grooved wheel upon
it, and is used in its simple form for the purposes of
hoisting alone. A rope fixed at one end to the circumference of the wheel, and wound about it, is unwound
by pulling at the loose end. The weight is attached to
the axle, or shaft, at the end of another rope, which
latter is wound round the axle (so raising the weight
and accomplishing its object) as that rope on the wheel
is unwound.

The gain in power is proportionate to the diameter
of the wheel in comparison with that of the axle, the
actual force exerted at any moment at the circumference
being, as will be readily seen, identical with that of a lever
having its fulcrum at the centre of the wheel circle, and its
weight attached at the circumference of the axle. This
power, though not much used in its simple form, exemplifies
the principle involved in many gearing wheels and pulleys.

F

The Pulley, which as a mechanical power must not be confounded with the pulley or rigger used on shafting, is a grooved wheel turning about an axis bearing in the pulley-block. A fixed pulley is only used to change the direction of a power, having no mechanical advantage in itself. Loose pulleys, however, when in combination with fixed pulleys, are sources of considerable advantage. The tackle mostly used nowadays for hoisting is a combination of two sheaves or blocks of pulleys, of which one is fixed and the other loose; the cord or chain is fastened at one end to the *fixed* block, and the other passed round, first one of the pulleys in the loose block, then in the fixed, and so on round as many pulleys as the sheaves contain, both of course having the same number. The power is then applied at the free end of this chain or cord, the weight being attached to the loose block itself. To reckon the mechanical advantage gained in this tackle, all that has to be done is to divide the weight by the number of cords or chains between the two blocks, not counting the end you hold in your hand.

The Inclined Plane as a mechanical power is not found in ordinary machinery, though its use for raising heavy objects is familiar enough in ordinary experience.

The Wedge.—This is really a double inclined plane, and is used for the sake of the increased lateral pressure exerted by its two sides, when driven by blows or pressure exerted in the direction of its axis.

The Screw is a circular cylinder, having a spiral thread running obliquely around its circumference. This works in a socket of its own diameter, having a precisely similar thread or worm cut in it to receive the screw; these are termed male and female, or companion screws. The power exerted by the screw at any given

point is precisely that of an inclined plane of the same
angle as the worm of the screw. The relation between
the screw and the inclined plane is precisely similar to
that which exists between the *lever* and the wheel and
axle. The power can be further augmented by turning
the screw, by means of a lever fixed into it at right
angles to its axis.

With this general acquaintance with the mechanical
powers, from the development of which all machinery
has sprung, we come to the consideration of the various
motions which are met with in an ordinary printing
machine-room. And the most logical, as well as the
most convenient method of dealing with them, would
seem to be to follow the power from the engine-room
to the delivery table, and notice, as nearly in order as
we can, the different changes it passes through in imparting its force to the various working parts of the
machines.

The power, then, by which a machine is run is delivered in a circular form from the driving rigger, or sometimes the fly-wheel of the engine—though this latter is to
be avoided if possible—and transmitted by means of a
belt or belts to the machine-room shafting, which, while
the engine is in motion, is continually revolving at a
regular speed. Upon this shafting pulleys, or riggers, or
drums—for they are variously named—are keyed immediately in line with each machine, and belts of leather
generally—though sometimes of other material—carry the
motion to them.

Before, however, we come to the machines, we must
devote a few remarks to the consideration of the pulley:
not, this time, the mechanical power so called, but the
ordinary machine pulley, and in connection with the belt

Pulley.—The pulley is a disc of metal revolving with the

F 2

shaft which carries it, and connected by a belt, strap, or band, with a similar pulley upon the machine itself. In the smaller pulleys, connection between the two is generally made by a cord or piece of gut, or rounded leather, fastened by a hook and eye to make it continuous. This method of connection, which is confined generally to small machines or parts of machines, necessitates a groove in the pulley, and makes the removal of the band a matter of some difficulty. As, however, in working large machinery, besides the requirement of transmitting a greater amount of power, it becomes at times a matter of imperative necessity to remove the band rapidly from the pulley, a broad-faced disc is used, upon which a wide leather belt travels. There is no fear of the belt slipping from the surface of the pulley, if, as is the case, this is made slightly convex. It might be supposed that the opposite construction would have best tended to keep the belt in position : but, on the contrary, the concave pulley would tend directly to drive the belt off its surface, the tendency of the belt in motion being to move upwards towards the centre. This may be seen by experimenting with a belt upon a cone-shaped pulley, when the effort of the belt to take a shape parallel to the surface of the cone drives it up, as far as its length will allow, to the thickest part of the pulley. It is upon this principle that the convex surface, which is, in fact, a double cone slightly rounded at the point of intersection, retains the belt, which, whenever it gets thrown too much to one side, is forced by the shape of the surface to return to the centre.

A belt passing straight round two pulleys drives the second, of course, in the same direction as the first. (Fig. 2.) Should it, however, be required to drive the opposite way, it is only necessary to cross the belt between the pulleys, when the motion will be reversed. (Fig. 3.) This principle is made use of in many machines for the purpose of reversing their action.

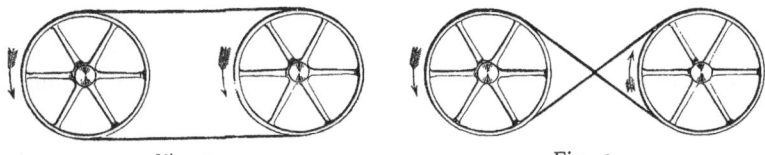

Fig. 2.   Fig. 3.

Fast and loose pulleys are fixed to every machine for the purpose of starting and stopping it at will, without the necessity of throwing off the belt. They consist of two pulleys placed side by side, of which one is keyed to the shaft which drives the machine, while the other rides loose upon it. The belt which is driven by the drum upon the driving shaft, is thrown by means of a fork from one to the other. Whilst it is upon the loose pulley, the latter runs free upon its shaft, without imparting any motion to the machine but when shifted on to the one keyed to the shaft, the machine is at once set in motion. It should be pointed out that the fork which shifts the belt is made to bear always against the advancing side of the belt. The retreating side of any belt or band may be pulled to one side at will without causing it to leave the rigger, *but any disturbance of the advancing side shifts it at once.* This fact is made use of in driving an axis which is not parallel to the main shafting, the only condition necessary being, that the advancing portion of the strap must always lie in the plane of its own pulley, the retreating portion being, in the case of two pulleys set obliquely, left to find its own way back without any fear of its leaving either pulley. If the motion of a pair of pulleys thus situated be reversed the strap will instantly leave them.

It is perhaps worth while pointing out that, by a well-known principle of dynamics, the strain on the strap is in inverse proportion to the velocity with which it travels. Hence, in transmitting power over great distances, it ·is necessary to drive the band at a high rate of speed, this being subsequently reduced by suitable mechanism, and the power gained where the machine is better able to stand the strain.

It is worth while to dwell for a moment upon this point, in order that the machinist may appreciate what is an apparent contradiction, but is nevertheless an important mechanical truth : that the strain on the belt or band diminishes directly in proportion to the velocity, and that a frequent cause of belts breaking is the attempt to run them at too low a rate of speed.

Suppose, for example, a belt is required to transmit 5 horse-power from engine to machine. Now, the work done by 5 horse-power in one minute is five times 33,000 foot-pounds, and the work done by the belt must be the same. Now, if the belt travels 600 feet per minute, the strain upon it is, in fact, one-six-hundredth part of the work done in that time, *i.e.*, 275 pounds. If it travels only 300 feet, the strain is double ; if the velocity be increased to 3,000 feet per minute, the strain would be correspondingly reduced to 55 pounds.

This fact has been utilised in what is called the teledynamic transmission of power, by which turbines or water-wheels are connected with machinery at a considerable distance, the power being transmitted by means of a slender wire rope or cable, moving at a very high velocity, and by its speed capable of conveying a high power, with proportionately a very small strain upon the cable.

At Schaffhausen, on the Rhine, three vertical turbines develop together about 750 horse-power, from which no less than seventeen factories in different positions are supplied with motive power transmitted by a single wire rope three-quarters of an inch in diameter, over a distance of some three-quarters of a mile. The linear velocity of the rope, which alone enables it to transmit so large an amount of power over so great a distance, is about 53 miles an hour.

At the London and North-Western Railway works at Crewe a traversing crane capable of lifting a weight of 25 tons is driven by power at the other end of the workshop, the transmitting medium being a cotton rope three-eighths of an inch in diameter, travelling with a velocity of 5,000 feet per minute. In this case the high velocity obtained is of course reduced at the receiving end, and the consequent gain in power utilised to move the crane without subjecting the cord to a higher tension strain than 109 lbs. It is curious to think of a three-eighths rope actually *lifting* in this manner a weight of 25 tons without breaking.

Gearing.—In the pulley just discussed we have an instance of the transmission of circular motion from one wheel to another. The principle upon which such transmission takes place is seen in its simplest form in the case of two wheels or cylinders placed in contact and turning upon one another. As, however, two such wheels (commonly known as friction wheels) are apt to skid in turning round when it is required to overcome any amount of resistance or transmit any very considerable power, it is necessary to provide them with teeth which fit into one another, thus insuring the perfect transmission of the energy from one to the other. For all that, two such wheels are considered for mechanical purposes as two circles rolling one upon another, and such circles are in fact conceived as forming what are called the pitch circles or pitch lines of the wheels, which determine their value in transmitting power.

Gearing forms a most important element in all machinery, and being very economical of space, is used in its various forms whenever circular motion of any magnitude is to be transmitted from one portion of a machine to another. It is not necessary that the wheels, in order to do their work, should have their axes parallel; it being a geometrical fact that any two right cones with a common vertex will roll regularly upon one another, and, further, that one of the cones may be flattened out into a level surface—such as a rack—without destroying this property. Hence it is only necessary to construct the bevelled edges of the wheel in such a manner as to fulfil this requirement, and, accordingly, gearing wheels may be placed in contact at any angle, if their teeth be properly cut. It is this property which simplifies the transmission of motion from one part of a machine to another, when the second axis of motion is not only not parallel to the first, but not even in the same plane, it being necessary only to proceed by successive pairs of suitably shaped wheels, changing the axis of rotation with each, until the required direction is obtained.

Broad wheels, having axes at right angles, are generally known as *mitre wheels; spur wheels* are those in which the teeth project radially ; *face wheels*, which have cogs or pins, to take the place of teeth, fastened perpendicularly upon the face of the wheel ; *annular wheels*, which have the teeth placed upon a ring inside the circumference of the wheel itself ; and *crown wheels*, in which the teeth are cut upon the edge of a circular band. All these are generally known as *gearing wheels*, the term *gearing* being in fact generally applied to all parts of a machine which are in close contact, and to all applied for a common object ; the expressions "*in gear*," and "*out of gear*," being respectively used (1) to indicate the position in which wheels, &c., are in connection, and capable of acting one on another or, (2) when they are shifted into a position where the teeth fail to catch.

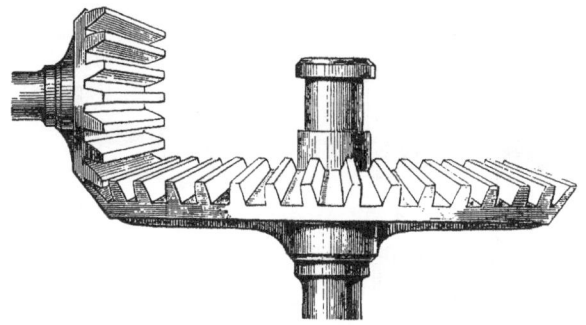

Fig. 4. --Bevel or Mitre Wheels.

The Crank.—Both the pulleys and the gearing wheels already spoken of are useful, as we have seen, only for the transmission of circular motion from one axis to another. One of the earliest problems in machinery, however, was the conversion of circular into reciprocal motion, and *vice versâ*, and this problem was eventually solved by the introduction of the *crank*.

A crank is simply a lever or bar, with a fulcrum at one end, and capable of being turned round by a force applied at the other ; being, in fact, a sort of link between the lever proper and the wheel-and-axle, of whose relation we have already spoken. By itself it has been long familiar as the ordinary handle by which a wheel is turned. It is in conjunction with the

connecting-rod that it assumes so important a place in
machinery. It is evident that if a circular motion is
given to the crank while the connecting-rod is hinged
to a straight bar travelling between girders or slides,
this latter will travel backwards and forwards as the
crank revolves, thus at once obtaining a reciprocating
motion from a circular one. The adoption of this principle was the first real step in the practical use of the
steam-engine, as it enabled a circular motion to be

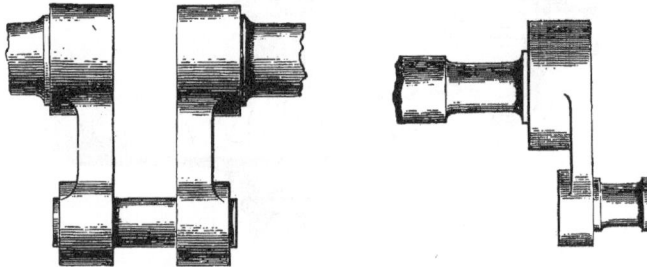

Fig. 5 —Cranks.

derived from the beam engine, which, up to that time,
could only be used for pumping purposes ; and by
applying the vertical action, already obtained, to the
turning of a fly-wheel, at once opened possibilities which
resulted in the various forms of modern machinery
The beam engine, which is used in many printing
offices to-day, would be of no value at all but for the
crank, which enables its motion to be resolved into the
circular motion of the shafting and pulleys. A simple
and even better known example of the crank is seen in
the ordinary horizontal engine—whether gas or steam—
in which the crank is the follower, and the motion is
converted from the reciprocal action of the piston to
the circular motion of the wheel.

Cranks are used generally in printing machinery for the reverse purpose of converting the circular motion, imparted by the driving shaft, into the backward and forward motion of the coffin and other parts of the machine. This motion, however, is imparted in other ways, notably by the rack-and-pinion arrangement, by which nearly all the ordinary perfecting machines are driven in this country, and of which we shall speak later

The *eccentric circle* is often described as a substitute for the crank and link, and supplies, in fact, an easy way of obtaining the same motion. It is often treated,

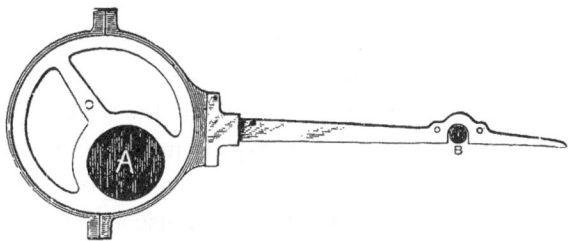

Fig. 6.—The Eccentric.
A Shaft ; B Stud ; C Eccentric.

however, as a form of the cam, of which we shall speak later. The simplest form of *eccentric* is represented in machinery by a circle revolving about a point away from its centre (whence the name), and surrounded by a ring or hoop, known as the eccentric band or strap, to which a rigid bar is attached in a line with the centre of the circle. It is evident that, as the shaft A revolves, the effect upon the eccentric will be to throw it, with the connecting-rod and stud B backwards and forwards, in precisely the same way as though the rod were keyed to a crank revolving in the same way about the centre of motion. The chief difference, and one which gives this form of motion an advantage over the crank and

connecting-rod, is that it does not require the rod to
travel so far out of the straight line, inasmuch as
the revolution of the eccentric gives the power of a
crank of the length between the centre of motion and
the centre of the eccentric, with a motion, or, as it is
called, a *throw* of double that distance. As this last
distance may be reduced at will (by bringing the two
centres closer together) *without reducing the power*, it is
quite obvious that, in such cases, a great advantage will
be found over the crank when a short motion only is
required. Another advantage, and an even greater one,
is found in the fact, which will be readily seen, that the
connecting-rod of the eccentric does not travel across
the centre of the shaft. Whenever it is required to
place a crank anywhere, except at the extremity of a
shaft, the backward and forward motion of the connecting-rod necessitates the sub-division of that shaft,
but the difficulty is avoided by the use of the eccentric,
which may be keyed upon the main shaft without interfering with its motion, as its own throw is entirely to
one side.

Cams.—We have spoken incidentally of the eccentric
circle being in fact allied to
the crank in principle and to
the *cam* in form. The usual
object of this latter contrivance
is the opposite of those we
have already discussed—the
conversion, namely, of circular
into reciprocating motion.

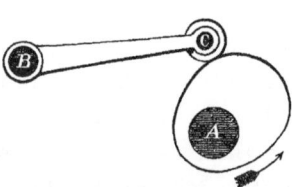

Fig. 7. —The Cam.

The term *cam* is applied to any curved plate which,
by the motion of its curved edge, communicates motion
to another piece. In the figure, we will suppose that
B is the end of a bar fixed in its bearings. The shaft A

has attached an irregular shape or cam, upon which the pulley C rests. As the shaft A rotates a peculiar motion is communicated to B by the rise and fall of C. Some cams are simple discs, with a *drop* in the flange, as in the motion which regulates the entry of the sheets into the drop-bar machine.

The horizontal machine of Applegath and Cowper's, which we described in Chapter I., first made use of this simple cam to regulate the entry of the sheet, imparting by its fall the necessary motion to the drop-bar which started the sheet into the tapes. The great advantage of this method of producing intermittent motion is its extreme accuracy.

Two other forms of *cams*, as applied to printing machinery, may be usefully studied—contrivances by which an interval of rest is obtained at the end of an impression by means of a cam-plate or groove. The first, devised many years ago for imparting a reciprocating motion to a *frisket* frame of the original Napier platen machine, consists of a combination of levers acting upon the frame itself. The end lever carries a sliding pin of an elongated form, which travels in the grooves of a cam-plate, consisting of two circular grooves, one outside the other. When the pin is travelling in one of the circles, the levers and frame remain stationary, changing into a position which moves the table the necessary distance, when the pin enters the other circle, and the frame remains again stationary during a second revolution of the cam. In the modern platen machine, the same motion is obtained in a simpler form, by the introduction of a revolving iron drum, upon which a helical groove is traced. In order to obtain a continuous motion, the groove is cut in the form of a right- and left-handed screw-thread, terminating at each end of the barrel in a flat ring. While the pin is traversing the groove from end to end, its motion is uniform, backwards or forwards. When, however, it reaches the circular ring at either end, it remains stationary during one revolution of the barrel. This peculiar motion will be more fully described in the chapter devoted to platen machines.

The instances given by no means exhaust the various applications of the principle of the cam, which any machine-minder may notice for himself. There is, in fact, scarcely any limit to the variety of motions which can be, and indeed are, produced by a suitable shaping of the cam-plate, a notable instance of which was given by the late Professor Cowper, who, in a lecture before the Royal Institution, arranged a combination which, when set in motion, described mechanically the initials R.I. by means of a pencil attached to one of a series of levers set in motion by a double cam.

Levers.—The conversion of circular into reciprocating motion by the crank and eccentric leads us naturally to speak of what are at once the simplest and the most useful of the working parts of all machinery, and particularly of printing machinery—the various kinds of levers which form so important a part in their construction. The arrangement of levers in machinery is usually known as link work, and the combination of two, three, or more levers, enables a motion once obtained to be transferred to other parts of the machine, while they can also of course be used to reverse, diminish, or increase the motion imparted to them in a hundred ways, which will be suggested to any one who studies a combination of this work in any machine, such, for example, as that given in the observations devoted to *cams.*

Levers are, for the most part, used simply to change

direction, as we have said. The simplest form of lever, indeed, appears in the handle by which the machine is started, which is, in fact, a plain lever whose power is used to shift the belt by means of a fork attached to it. What has already been said of the mechanical attributes of the lever may be applied to any particular instance, remembering only that motion through increased space is accompanied by loss of power, and, *vice versâ*, that power is gained only at the expense of space, a rule which holds good in all mechanical action, but is especially easy to notice in the case of the lever.

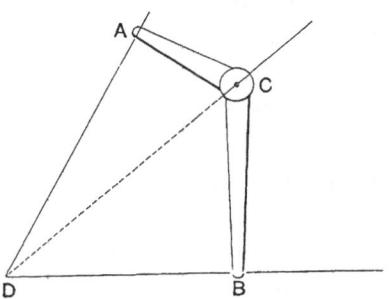

Fig. 8. —Bell-crank Lever.

Several special combinations of levers have particular names, and may be spared a few special words. The *Bell-crank* lever, so called (very largely adopted in printing machines), is simply a bent lever, whose object is to change the line of direction of some small motion. It consists of two arms standing out from a fixed axis. It is apparent that when the point C at the end of the arm B C is moved slightly in the direction C D, the point A moves slightly in the direction D A, and *vice versâ*. The shape of this lever necessarily varies, both the length of the arms and the angle being arranged to suit requirements. The Bell-crank is only adapted for light work, such as vibrating rollers, drop-bars, &c. It enters largely into the construction of the simpler motions of the Wharfedale machine.

The *knuckle-joint*, or *toggle-joint*, as it is commonly called, is only another application of the same principle.

It consists of two arms jointed together and fixed at one end only, the lever end being connected with a guided rod, which confines its motion to the line of the other. Any one who has ever tried to straighten a long wire or heavy cord, must have realised that the force necessary to straighten it increases as it more nearly approaches a straight line ; and a really heavy cable, like those of a suspension bridge, cannot be straightened, as it would break first. The same principle reversed shows at once the power of the knuckle-joint, which is accordingly employed in the Napier platen machines to depress the platen, the power increasing as the platen descends, and being greatest at the moment of impression. In this case the knuckle-joint is attached to a revolving crank or eccentric, by means of a connecting-rod, the revolution of the crank straightening the joint once in every revolution, and then opening it again in order to draw back the platen after the impression has been received.

Racks.—The possibility of changing the direction of a movement through a greater or less angle has been already discussed under the head of gearing. A few words, however, may be added as to the special manner of driving adapted in the perfecting machine, which depends for the motion of its tables upon a rack driven by means of an upright spindle. The motion of this and of the *rack* which is propelled by it may be conveniently taken together, and both best explained by taking the actual working of the machine, as familiar, in appearance at least, to most printers.

The tables themselves run between two parallel frames, and rest upon a set of pulleys or runners on either side of the machine, which support them as they travel to and fro. They are driven backwards and

forwards by the motion of a *rack*, which is fitted underneath them in such a manner as to permit of its shifting
its position from one side of the machine to the other,
at the end of each motion. This rack in turn is driven
by an upright spindle, which is driven at right angles
by a bevel wheel turning upon a horizontal driving shaft.
To secure strength, this spindle is fixed immediately
underneath the centre of the machine, and is secured in
the middle by a bearing, while its base rests in a socket
in the bed of the machine, as we shall fully explain
in Chapter VIII.

Another form of rack is composed of a series of cogs,
arranged in a straight line. These are used on either
side of the Wharfedale, and work in gear with the
cylinder.

The *Mangle Rack* of the Anglo-French machine is
different from both the preceding. The teeth of the ordinary rack are inverted, so to speak, the base and sides
of the cogs being free, so as to admit of their being geared
to the pinion of the upright spindle. The cogs of the
rack of the Wharfedale assume a perpendicular position,
that the cylinder wheel may work in gear from above ;
while the teeth of the mangle rack are fixed into the
side of a bar, and lie free, in a horizontal position, to
allow the pinion to work in gear both on top and
underneath.

Parallel Motion.—Parallel motion, as used in steam-
engines, was first designed by James Watt, who described it in the specifications of a patent he took out in
1784. It was intended particularly to enable the end of
the beam of a beam engine to move the piston-rod of a
pump vertically up and down, without twisting or bending
it by the curve of its own motion.

The application of this principle to the motion of the

G

rack is exemplified in the ordinary perfecting machine
(see Chapter VIII.).

Universal Joint.—The universal joint, invented by Dr.
Hooke, is a contrivance for allowing two shafts to have
perfect freedom of action in any direction within certain
limits.   It consists of a hollow square of metal, to alternate sides of which the forks in which the shafts terminate are loosely bolted.

This motion is adopted in the Anglo = French
machines in which—or in the majority of them—the
mangle rack is driven by a pinion, which, in lieu of the
upright spindle and bevel wheel, is attached by means
of a universal joint to the driving shaft itself, the joint
enabling the pinion wheel to change position from side
to side of the rack, which is fixed, instead of requiring
the rack itself to shift across the bed of the machine.

Segment wheels were amongst the chief features of
Mr. Cowper's first printing machine, in which the sheets
were fed-on by the intermittent engagement of a wheel
with a small sector working a drum and continuous series
of tapes, the motion being given to the sheet only when
the two met.

Another form of intermittent motion commonly employed in machinery is the *ratchet,* mostly used for rotating
the ductor cylinder.   This name is given to a wheel with
teeth of a suitable form, which is driven by a vibrating
piece.   This piece or paul works loosely upon the end of
an arm or lever, and when propelled forward moves the
ratchet a regulated distance.   When returning, the paul
slips over the points of the teeth and engages a fresh
one on the arm, returning again in a forward direction.

As before mentioned, the ratchet is adopted in many
machines on the end of the ductor roller, which is moved
forward in this manner intermittently, by which means

the supply of ink to the vibrator may be conveniently regulated.

Reversing Motion.—The motion of an engine may of course be reversed at once by admitting the steam into the other end of the cylinder. The ordinary motion, however, of the machine rooms is derived from the continuous motion of a main shaft, from which it is taken off by pulleys and belts to the different machines. It would obviously be impracticable to reverse the motion of this shaft in order to run back a machine, and it is necessary, therefore, to notice the mechanical means which we have of reversing the motion in the machine itself. The first and one of the commonest instances of reversing motion, which we give here as an example of those motions which depend upon the use of gearing wheels, is derived from the simple combination of two and three wheels respectively.

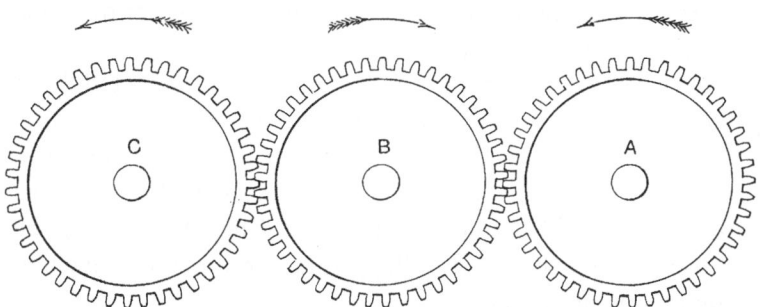

Fig. 9.—Reversing Motion.

If A B C represent three wheels in gear with one another, it will be seen at a glance that A and B turn in opposite directions, while A and C turn the same way. It will be clear, then, that if A represent the direction of motion of the main shaft, the motion will be transmitted direct by gearing the three wheels together

G 2

to the further wheel, while, if two wheels only are used, the second wheel will revolve in the opposite direction and the motion be reversed. If, then, we connect two parallel axes with a combination of two and three spur wheels alternately, we have an example of reversing motion very constantly used.

Various modifications of this principle are to be found in different classes of machinery, but they suffer from one of two disadvantages, having either to drive an unnecessary number of wheels, thus increasing the friction, or depending for their action upon the shifting of other wheels in and out of gear, thus interfering with the accuracy of the machine. A simpler and better plan for obtaining a reversing motion which will be more familiar to most machine-minders, is that which depends upon the crossing of the bands. It has been already pointed out that by a band conveying motion between two pulleys, one of them may be made to revolve in an opposite direction to the other.

The reversing of the rolling machine is effected by a pair of bands running from the main shaft in opposite directions (one crossed and the other straight) Between the live pulleys is the driving pulley, keyed on to the driving shaft. By means of forks either band may be directed upon the fixed pulley, while the other is travelling round the live pulley.

The motions which have been described in the preceding pages are of such a general character as may be found in almost all machinery, and an understanding of the principles upon which they are constructed and the method of their working may enable the workman to enter upon the study of the particular way in which each of them is applied in the various types of machine which may come under his notice. Some of them have been already touched upon, and we shall shortly come to the examination in detail of each of the principal machines in use in the trade. To this time may be left those more complicated combinations and modifications of the simple motions which we have described, as they occur in practice. It may be worth while, however, to speak of one or two of these special motions before quitting this branch of the subject altogether.

Such a combination is the *rocking frame* of the Anglo-French machine already spoken of. In this machine the large cylinder used in the Web and Drop-bar machines is replaced by two smaller cylinders of about half the diameter. Nearly all the surface is used for the impression, as the circumference of the cylinder is but slightly in excess of the surface of a full-sized forme. As there is, therefore, no idle space under which the forme can return, it is necessary for this purpose that the cylinders should be alternately lifted from the level of the type in order that the forme may travel back and clear the cylinder. This alternate rise and fall of the cylinders is effected by means of a rocking frame in conjunction with knuckle-joints or levers, as described in Chapter IX.

The term "rocking cylinder," with which the above must not be confused, is applied to the cylinder of what is generally known as the "Tumbler" machine. This machine—also called the "Main," after the inventor—was the first successful single-cylinder machine, and its peculiarity, as will be explained later, consists in the rocking action of the cylinder, which, instead of revolving completely, and remaining stationary, "tumbles" back, as the forme returns, to be ready for the next impression. At the same time it is raised slightly from the coffin or table to allow the forme to return, as we have seen was necessary in the Anglo-French machines. The rocking here is effected by a bell-crank, as will be more fully explained in the chapter descriptive of the Main machine.

This "tumbler," sometimes also called a "pulley," is practically a small wheel, which acts as a guide to some larger wheel or combination of wheels, running itself in most cases over a cam-plate, and by "tumbling" or changing its position as it reaches indentations or changes of curve in the plate, determining the change in motion of the whole combination. Tumblers are found in all parts of the printing machine serving such purposes as these, and several have previously been referred to in describing other motions.

Springs are also used freely in combination with other motions, their use being generally to assist an action or motion already commenced by some mechanical means. Thus the rocking frame of the Anglo-French machine already spoken of, is merely steadied and kept up to its work by the spring between the cylinders, its actual motion being dependent upon the mechanical action of the working parts of the machine. Springs are used to close the grippers on all machines where this action is used. In the Anglo-French machine two strong springs, placed horizontally, are usually fixed so as to receive the tables at the end of their travel, and ease the sheet on their return. Every machine-minder will be able to find a number of instances in which such springs are used. The usual form of them is that of a coil of stout wire inside a cylinder, but other forms of springs are occasionally employed.

Brakes are fitted both to the fly-wheel and to the cylinder wheel of the Wharfedale, Bremner, and small platen machines, in order to insure steadying of the cylinder and the prompt stopping of the machine. The brake is set in action by a lever resting upon a cam, cut so as to distribute the pressure, which is accordingly unequal, being strongest when it is required to overcome the strain of the impression. The strikers of several machines are connected with a fly-wheel brake which acts automatically upon striking off, and stops the machine instantly.

Balance Weights.—In the Napier platen, in order to assist the platen in rising after the impression, and to steady it generally, a heavy weight extends across the machine, and is fastened by means of gut bands to the chills of the knuckle-joint. Balance weights are also used to draw back the marks to their places in the Web machine we have already spoken of.

In the foregoing pages an endeavour has been made to describe the chief mechanical motions which are employed in printing machinery. In the description of the various machines, all or most of these will be found to play an important part in the mechanism. Meanwhile a glance backward to the history of the invention and growth of the printing machine described in Chapter I., will show very clearly the development of these various forms of power and mechanical motions as one after the other the knowledge of them which mechanics had acquired came to be applied to the machine, until it reached somewhat of the perfection of the modern steam press. It is to be hoped that the reader who has followed us so far will be able to bring to the study of the various types of modern printing machinery with which we shall be concerned in the sequel these two ideas at least : that

the machine did not spring into being in its present
shape, but is, in effect, the product of a vast number of
successive improvements and developments, in which the
characteristics of the original invention are almost lost sight
of in the changes which have passed over it; and, secondly,
that, complicated as any given piece of machinery may
appear at first sight—and as in fact it may actually be when
regarded as a whole—yet each and every machine is composed only of those simple and easily understood motions
and working parts with which we ought by this time
to be familiar, and that to understand it, it needs only
a proper analysis and resolving of its complications
into the simple motions which compose them. With
such a grasp of the subject we may safely start upon the
detailed consideration of some of the principal printing
machines in use at the present day.

CHAPTER III.

ONE-SIDED STOPPING CYLINDER MACHINES.

Description of a typical one-sided Machine of modern Manufacture
—The Wharfedale: its Peculiarities and Advantages—Taking-
off Apparatus—The Bremner—*Graphic*—Dawson's—Payne's
— The Quadrant—The Ingle—Newsum's Anglo-American—
Furnival's—The Reliance—Cuthbertson's Patent Counter.

THE ordinary one-sided machine is usually known in
the trade as "The Wharfedale," whether or not it be
manufactured in the locality from which it derives its
name. Previously the single-cylinder machine mostly in
use was the "Tumbler," which gradually became superseded as the advantages of the stopping cylinder came
to be appreciated.

Although there are several machines, all more or less
excellent—each bearing a different name, and claiming
peculiar advantages for "patent" appliances—all are really
built upon the same principle, and may be said to
consist of an impression cylinder, mounted upon parallel
side-frames. The "coffin" or table with the ink-slab
moves to and fro, carrying the cylinder in gear one
revolution, when travelling outwards, and leaving the
cylinder stationary on returning, so as to admit of the
sheet being laid into the grippers for the next impression.

The cylinder is a large hollow drum, having an opening of about five inches wide on the under-side along its
length. It is secured upon the top of the side-frames in
gun-metal bearings, by strong brackets. At each end of
the shaft, inside the brackets, is a cog-wheel of the same
diameter as the cylinder, the one on the laying-on side

FIG. 10.—HARRILD AND SONS' REGISTERED BREMNER MACHINE.

being *fixed*, while the off-side wheel runs loosely to and fro with the table rack.

The coffin, or table, and the ink-slab are supported upon two or more girders resting upon the end frames, and an additional cross-stay immediately beneath the

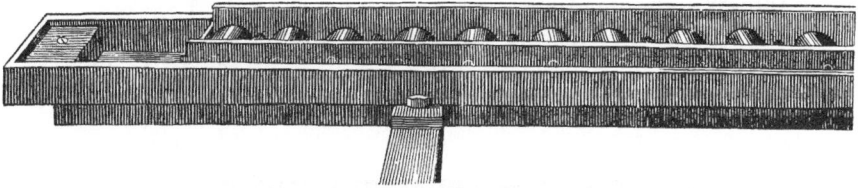

Fig. 11.—Showing Position of Runners in the Parallel Bars.

cylinder. These girders have flanges on either side, thus forming extended grooves. The width of these varies with the size of the machine, the larger ones being between three and four inches wide. Independent of both table and girders are a set of runners, which travel

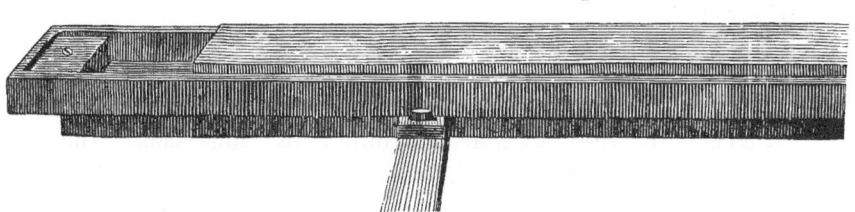

Fig. 12.—B Table-bar on Top of Runners.

freely in the grooves. These runners—a series of small substantial, solid steel wheels—are fixed at intervals between parallel iron bars, the latter fitting closely into the grooves or slides on top of the girders before mentioned.

Attached to the under side of the tables, and extending their whole length, are carefully planed steel bars, which drop immediately upon the runners, thus affording

firm support, and at the same time allowing perfect freedom in travelling to and fro.

To prevent the runners skidding, or one set moving
at a different speed from the other, they are sometimes
held together by a thin iron rod extending across from
one set to the other.

It will be readily understood that there is but a
slight strain exerted by the movement of the tables,
excepting at the point immediately beneath the cylinder

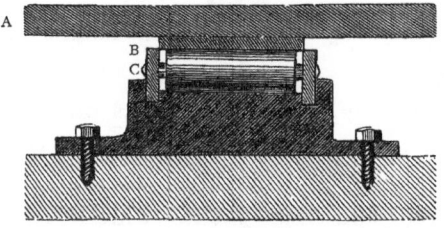

Fig. 13.—Section showing Table, Table-bar, and Runner.
A Table ; B Table-bar ; C Runner.

when the impression is being taken, it being only necessary to *support* the table at any other portion of the
travel.   In the original machines of this class, the
makers, singularly enough, seemed to have overlooked
this fact.   The result was, that the tables, unable to
offer proper and firm resistance at this part (especially
when the overlaying had been badly done), actually bent
under the pressure.   This necessitated the placing of
wrappers and glazeboards in the centre (even in the case
of movable forms), gradually levelling up towards either
side, when the forme was originally laid on.   This defect
was, however, gradually remedied, either by the fixing of
large independent pulleys immediately under the cylinder,
or by the addition of a central set of runners.   Most makers

now put a cross-girder in addition, thus affording a sufficiently firm resistance to the pressure exerted above.

The motion to the tables is imparted by a strong
steel connecting-rod fixed to a large geared wheel or
wheels, driven from a small cog on the main shaft. The
other end of the rod is connected to a short shaft having
large and substantial cog or traverse wheels on either side.
These wheels travel in parallel racks, one set on the under-
side of the table, and a corresponding pair at the base.

Although the tables may be said to be supported by
these large geared wheels, it must not be understood that
the teeth "bottom." On either side of the cog is a flange,
which runs upon a steel bar immediately at the side of
the rack, and upon a corresponding flange parallel with
the racks under the table. Thus the tables, while being
supported upon the flanges, are propelled by the gearing.

The driving wheel is fitted with a balance weight on
the opposite side to the connecting-shaft bearing, that
the strain or weight of the end of the rod may be to
some extent equalised. In some machines, the weight is
independent of the wheel, but of course placed relatively
in the same position.

On the smaller machines, a rocking lever is used in
lieu of gearing. The fixed point of the rocker is at the
base of the machine, the other extremity being attached
to the under-side of the table. The connecting-rod is
secured to the centre of the rocker, and moves the table
to and fro.

Every movement is strictly in gear, which necessarily
insures absolute steadiness while in motion. The most
careful attention is paid by the makers to the cutting of
the racks and wheels, as, if any play were allowed, slurring
would certainly result.

As before mentioned, the cylinder is stationary during

the travel of the table back towards the laying-on board. To enable the forme and cylinder racks to clear, a small portion of the under-side of the cylinder from the grippers is slightly flat, together with the fixed cog-wheel. On the return of the table, the cylinder is forced into gear with the racks on either side.

The grippers (Fig. 15) are arranged along a bar or

Fig. 14.—Ingle's Machine, showing Rocking-Lever Motion.

rod in the opening of the cylinder, and drop slightly lower than the front edge of the brass plate of the laying-on board, that the sheet may be placed into position flush against the front marks. The grippers, as will be seen from Fig. 16, are independent of one another, and fit more or less tightly along the gripper-bar or rod, each being provided with a small set-screw, in order that both its position and its pressure to the edge of the cylinder may be regulated to a nicety. The opening and shutting are accomplished by a slide and crutch on the off-side of the cylinder. The gripper-bar is carried outside the cylinder, and is provided with a short arm, at

the end of which is a solid pulley. When the pulley is
free, or is not in contact with anything, the grippers are
shut closely upon the edge of the
cylinder, being held by a substantial
spring inside (Fig. 16). The grippers
are opened for the reception of the
sheet by the action of a crutch,
attached to the end of a rod im-

Fig. 15.—The Gripper.

mediately beneath the side of the cylinder, which rises
and forces the tumbler upwards and the gripper-bar down.
Directly before the return of the table for the impression,
the crutch drops, and the spring inside the cylinder pulls
the whole of the grippers flush to the edge of the
cylinder, and secures the sheet. When the impression
is completed, the pulley runs on to a slide, which again
opens the grippers, and releases the newly printed sheet
enabling the boy to take it off. The slide referred to
extends about six inches, of course following the shape of
the cylinder, and at its base joins the movable crutch
before mentioned—so that when it is at a standstill, the

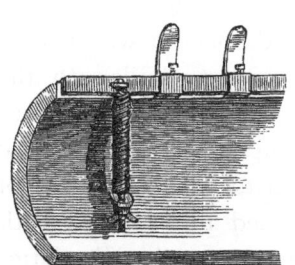

Fig. 16.—Section of Cylinder,
showing Gripper-bar and Spring.

grippers are open, the crutch
having returned (immediately
after the cylinder was forced
into gear) into its original position. When the crutch, therefore,
is raised flush to the level of the
fixed slide, the grippers are open,
the pulley being forced in an
upward direction. When the
crutch drops, releasing the pulley,

the spring inside the cylinder forces the grippers close to
the edge of the entry. On some machines the side-slide
and crutch are made in one piece, but the action of the
gripper-bar is not affected by this modification.

The crutch is worked from a large cam on the cam shaft at end of machine—the same as that to which the connecting-rod wheel or wheels are attached. The cam wheel, therefore, makes one complete revolution to every travel to and fro of the tables. A steel rod with a pulley at the end is so supported as to press against the shape or cam, and, being attached at the other extremity to the base of the crutch by a bell-crank, imparts to it an up-and-down motion.

As before stated, at either side of the cylinder, inside the frames, is a toothed wheel, of same diameter as the cylinder itself. The one on the off-side is fixed to the shaft, flush with the cylinder, and is therefore practically a part of the cylinder itself—its function being to steady the working and equalise the strain when the impression is being taken. The loose wheel on the other end, or near-side, is for the purpose of forcing the cylinder into gear with the table racks. On the inside, at the base, is secured a stout steel shape, working at one end upon a pin, and held in position by a spring. This is attached to a small tumbler on the other side of the wheel, and, by means of a small crutch, somewhat similar to the one already described, may be raised or lowered. Immediately after the grippers close upon the sheet, on the return of the tables towards the ductor, the loose end of the shape is raised, and is thus forced into contact with a steel stop on the end of the cylinder, giving the latter the necessary impetus to carry it into gear with the table racks. The cylinder then makes one revolution and stops, the tables returning, together with the wheel, which runs back loosely on the shaft.

When the boy lays a sheet badly, or for any other reason desires to prevent the cylinder from travelling, he simply strikes off the handle governing this motion. The bar working from the cam shaft is thereby lifted from a pin on the upright crutch rod, and the latter remains stationary, allowing the tables to travel forward without the cylinder being forced into gear. By an ingenious arrangement, the cams on the shaft can be so adjusted that the crutch may be struck off automatically at every other impression, thus obtaining double inking. To prevent the cylinder being struck off at the wrong time—when it is practically in gear—a wheel of the same diameter as the cylinder is placed on the extreme end of the shaft on the near-side, outside the side-frame. At the bottom is a slot, connected with the striking-off motion, and inasmuch as it is impossible to strike off unless the small shape is allowed to fit directly into the slot, the striker-off cannot be used excepting the cylinder be absolutely stationary, as the shape merely presses on the level flange of the wheel, and cannot be forced down until the latter comes exactly into position.

It will be readily understood that success in printing is largely dependent upon the construction and gearing of the cylinder. If its relative position changed with each impression only to the extent of a thin lead (about $\frac{1}{45}$ of an inch), the overlays would be thrown off and the sheet spoiled. To insure its being exactly in the same position every time between the impressions, a push-back bar is used. This consists of a steel rod working on a cam at the end of the cam shaft on the off-side, outside the frame. When the cylinder stops, it slightly overshoots itself. The push-back bar, however, immediately moves against a steel shape on the bottom of the brake wheel, and gently forces it back exactly into its correct position, the other side of the slot being jammed against a stout steel bell-crank attached to a spiral spring. When the cylinder is completing its revolution, it forces the bell-

H

crank down, but the latter immediately returns to a horizontal position, and is unyielding when pushed in the same direction.

An automatic brake is also applied to the cylinder to ease its return and prevent a jar, which might be caused by a too sudden stoppage. The wheel above mentioned has a broad flange. By means of a rod working from the cam shaft, a small brake, made of

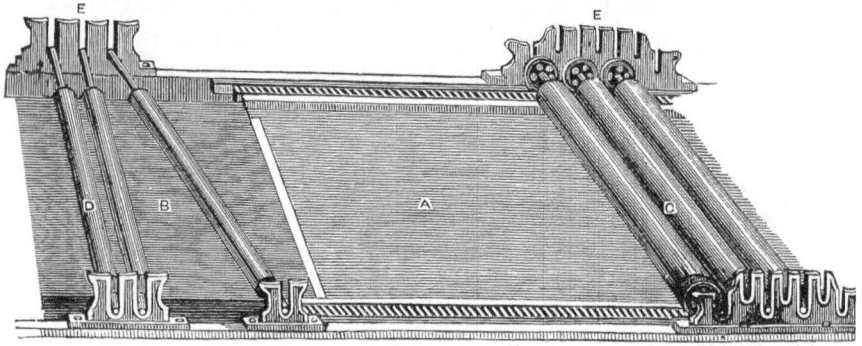

Fig. 17.—Coffin and Inking Slab, showing Arrangement of Rollers.
A Coffin; B Inking slab; C Inker; D Waver; E E Roller forks.

wood, lined with leather, is forced against the flange immediately before the cylinder completes its revolution. The brake drops when the cylinder stops, to allow the push-back bar to perform its work.

The inking arrangements of this class of machine are excellent—as many as five inkers, in some cases, entirely covering the forme. The duct occupies the usual position at the extreme end, and the roller is worked by either the ratchet motion or a gut band. In the former case a small shaft extends outside the frame, to which is attached a solid wheel, having a slot cut from the edge to the centre. In this slot works a slide, attached to the arm extending to the geared wheel on the end of the ductor roller. By a set-screw the slide can be regulated to a nicety near the flange of the wheel, giving the arm greater travel, or near the axis, when the motion is very slight. The ratchet works loosely on its pin, and can be thrown back, throwing it out of gear, when desired. Its own weight keeps it down upon the teeth of the wheel. The vibrator is fixed into a rod, having bearers on either side. This is moved up and down by a cam on shaft immediately below.

Some machines are provided with an independent inking drum, parallel with the ductor, and having a slight motion from side to side, which assists the distribution. In this case, of course, two vibrators are necessary—one working horizontally from the ductor to the drum, and the other perpendicularly from the drum to the table. The same lever works both vibrators. The set- and draw-back screws of the ductor are arranged in the same manner as in other machines.

An ingenious set of lifts are also adopted by some makers, by which means the whole of the inkers may be lifted sufficiently high to enable the forme to travel without coming in contact. The treadle by which this supplementary frame is worked is at the base of the machine, and, assisted by balance weights, is easily worked. This addition is greatly appreciated by the machine-minder, as it frequently saves the trouble of lifting the inkers when making ready.

The laying-on board is made in two parts—the back portion upon which the white paper is placed being perfectly level upon the side-frame, and the front portion falling slightly towards the cylinder entry. When the paper is being fed-in, the front edge, usually faced with brass plate, is below the gripper-bar, the grippers, however, being underneath. Immediately before the

H 2

grippers close upon the sheet the front edge of the board is raised about an inch by a rod outside the frame pushing up a small plate which hangs slightly over the side. This, with several other motions on the machine, is brought into play by a bell-crank, and worked from the cam shaft. The brass gratings on the board are for pointing purposes. Underneath—*i.e.*, immediately below the board—is a steel skeleton frame, so constructed as to admit of the points being screwed in almost any position. When the board is down, the points appear *above* the grating, and the boy may place his sheet upon them. The raising of the table immediately before the grippers close, lifts the paper above the tops of the points, and allows the sheet to be removed by the grippers without impediment. On some machines the board is stationary. In this case the points fall below the top of the gratings and release the sheet. The boy, when pointing, stands immediately in front of the cylinder, the back portion of the board being removed, and the pile of work placed on a bank at the side to his right.

The smoothers are a set of brass fingers somewhat resembling the grippers, but fastened along a bar immediately in front of the cylinder on the front edge of the laying-on board. They are provided for the purpose of taking the crease out of the sheet when the paper happens to be in bad condition. These smoothers may be fixed as near the surface of the laying-on board as desired by means of set-screws. They are seldom used, we are glad to say. They are simply supplied to remedy a defect which of course should not exist, and are liable at times to cause much trouble.

When double inking is required, the tables are made to travel backwards and forwards *twice* between each revolution of the cylinder. This may be done, as before described, by the layer-on striking off the cylinder every other revolution, but the advantage of automatic action is obvious.

The time occupied by printing with double inking is, of course, twice that of inking singly in the ordinary way, the output of the machine being only one-half of its actual capability. To meet this objection "double-ended" machines have of late been constructed—a supplementary ductor with wavers and inkers being fitted at the laying-on extremity of the machine. This necessitates a slightly longer travel of the tables, but affects the actual speed of the machine very little. With this additional facility, however, the finest class of work may be done.

The Wharfedale machine is admirably and conveniently fitted, and every facility is afforded the printer for storing and resting the rollers. The taking-off board is provided with a set of racks on the under-side. When thrown up, to allow of the forme being attended to, the inkers can be lifted and lodged in these racks, thus obviating the necessity of taking them away from the machine. At the end of the machine, under the laying-on board, is also fitted a set of racks, which may be used for storage of rollers not required for immediate use.

In some cases the roller forks are made independently of the side-frames, and are easily secured by a couple of stout pins which fit in the flange of the roller-fork frame, and into a slot on the top of the side-frame Slight dips are also made between the forks, to admit of the inkers resting free from the forme when the taking-off board is down.

The strikers for setting the machine in motion and stopping the cylinder are at the side of the frame where the layer-on stands. When pointing (excepting when two boys are employed), the layer-on stands in front of the cylinder, the back laying-on board having been removed. A duplicate set of strikers, connected by bell-cranks, are provided for this position. On some machines a powerful brake is attached to the striker. This may be considered an advantage, as the fly-wheel is clamped, immediately the strap is removed, to the live pulley, and the machine is stopped instantly.

TAKING-OFF APPARATUS.

The automatic taking-off of the sheets is effected by means of a small drum on the top of the impression cylinder, and working in gear. This drum is half the diameter of the cylinder, and is provided with a set of grippers so arranged as to enter the opening of the former immediately the sheet is released, at the same time clutching and releasing it into a series of endless tapes on a frame slightly above the laying-on board. Lying between and slightly below the level of the tapes are a set of " flyers," *i.e.,* a series of wooden laths fitted into a stout bar at one end. Immediately the sheet is run on to the tapes, the flyers rapidly move in a semi-circular direction over to the taking-off board, depositing the sheet upon the heap, printed-side upwards. This movement is effected by a series of cogs working in a small wheel at the end of the bar holding the fingers, and derives its motion from a cam on the main shaft. Immediately the sheet is laid, the flyers are moved back ready for receiving the next sheet. To facilitate the travel of the newly printed sheet and to keep it strictly in position, a set of thick indiarubber rings are placed along a spindle immediately above and flush to the taking-off drum. These, having the motion imparted by the drum, pass the sheet on to the tape frames, the grippers having released the edge. The indiarubber rings may be placed in any position along their shaft, and thus may be arranged by the workman so as to

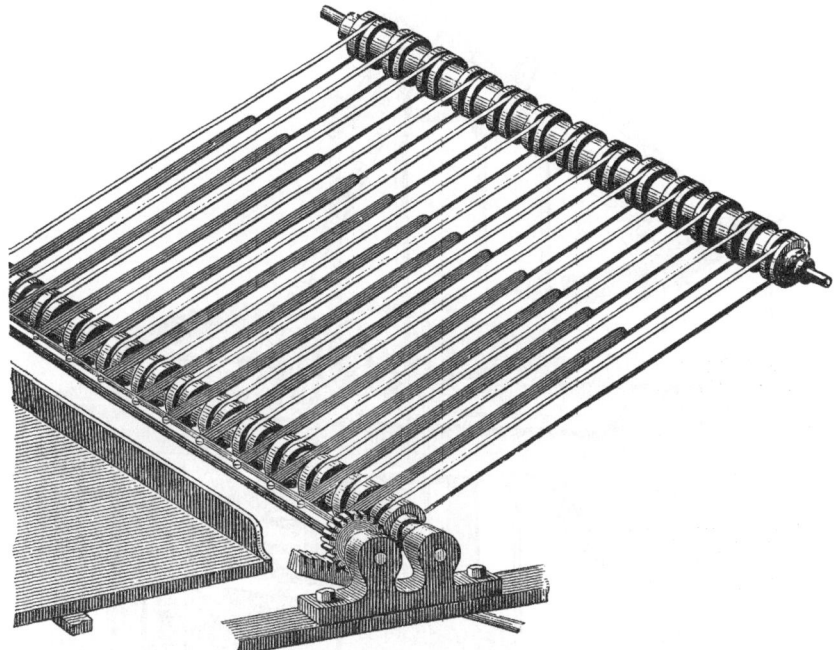

Fig. 18.—Flyer of Taking-off Apparatus.

fall into the gutters, or blank spaces between the pages, otherwise the newly printed matter would be smeared by the contact.

MARK SMITH'S TAKING-OFF APPARATUS

is certainly a most ingenious arrangement. It consists of four uprights fixed to the side-frame of the machine, immediately above the cylinder. On either side of the

cross-bars of these uprights—a small extended frame—
an arm is attached, and a second frame dipping down
towards the laying-on board, the length of which is

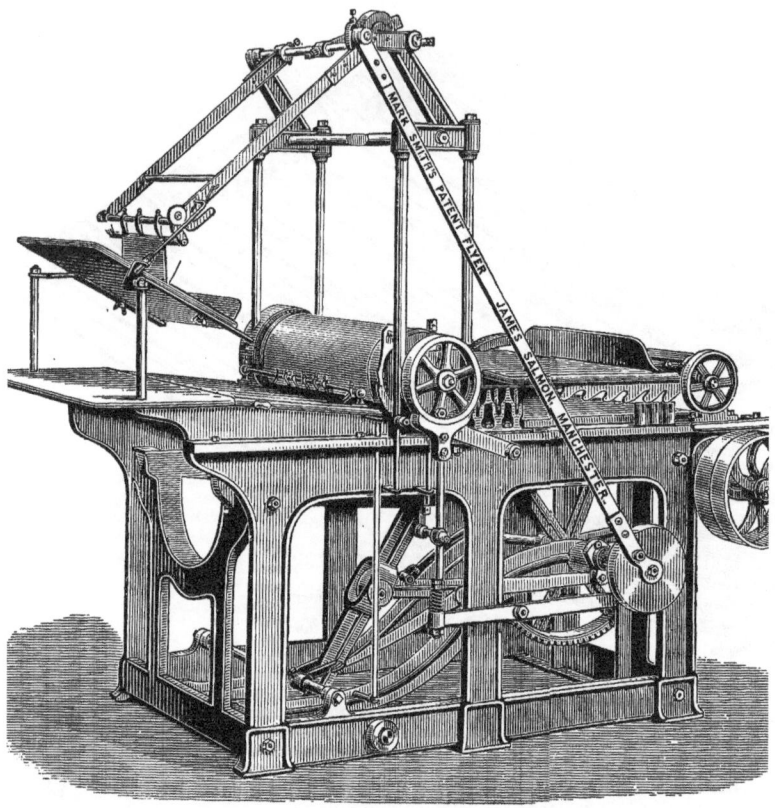

Fig. 19.—Small Jobbing Machine fitted with Mark Smith s
Taking-off Apparatus.

regulated by the size of the machine, having a series
of grippers arranged along a bar at the extremity. This
frame has a movement extending from the base of the
cylinder on the laying-on board to the taking-off board,
which is fixed on brackets at an inclined angle behind

and above the laying-on board. This gripper frame is supported and moved by a steel bar extending from a wheel on the cam shaft outside the side-frame. The grippers at the extremity are governed by a cam. When the machine is set in motion the frame moves downwards, and the grippers enter the cylinder, opening immediately upon the release of the sheet, grasping it and returning above the taking-off board, dropping the paper on the heap before again moving downwards. The work is laid with the newly printed side uppermost, and there is no probability of a smear or slur occurring through the wet ink coming into contact with tapes, &c., as the sheet is grasped along the lay edge. An advantage very justly claimed by the makers is, that the ordinary taking-off board may be removed entirely, giving the minder a full view of the forme, inkers, &c., during the working, if desired.

This apparatus, which has been largely adopted, was introduced in 1878 Unlike the taking-off flyers previously described, it may be conveniently fitted to the Tumbler machine. Messrs. Salmon, of Manchester, are the makers.

As before stated, there are many machines built upon the general principle described ; the Bremner, by Messrs. Harrild ; the Wharfedale, by Messrs. Dawson and Messrs. Payne ; the "*Graphic*," by Messrs. Parsons and Davis—besides those made by Messrs. Furnival, Messrs. Fieldhouse and Elliott, Messrs. Powell, Messrs. Newsum, Wood, and Dyson, and Messrs. Ingle.

THE BREMNER.

This machine is largely used for the production of the finest class of cut-work, and has an established reputation for general excellence. In appearance it is slightly more complicated in construction than others, but this is to a great extent owing to the provision of several minor conveniences, which, it must be admitted, enhance its value. It may be mentioned, among other small details, that by an ingenious arrangement, consisting of a cam worked in a boxed wheel, the brake on the fly-wheel may be acted upon at any point of its motion, so as to slacken the travel of the machine when on the impression. This is of great assistance to the colour-printer when a solid forme is being worked, as it is almost impossible to lift rapidly a sheet covered with a great body of ink without leaving a portion of the surface of the paper behind, unless, indeed, the ink be very thin, in which case the printing is unsatisfactory, the impression lacking "body." Directly the grippers release the sheet, the brake drops from the fly-wheel, and the table returns at the usual speed. This machine is most elaborately and splendidly fitted, and (although elegance may be said to contribute but little to general efficacy) as a specimen of handsome printing-engineering work it is unrivalled.

FRANCO-BREMNER MACHINE.

This was constructed upon the principle of the French maker Dutartre, whose machine was, we believe, the first really "double-ended" machine introduced into this country. Mr. Bremner, wisely admitting the advantages of increased inking power, designed a machine on the lines of the French maker, adding all the improvements embodied in those of latest construction. The laying-on board is filled at an angle similar to that of the litho machine. The sheet is laid into the grippers at the top instead of at the bottom of the cylinder, and released to

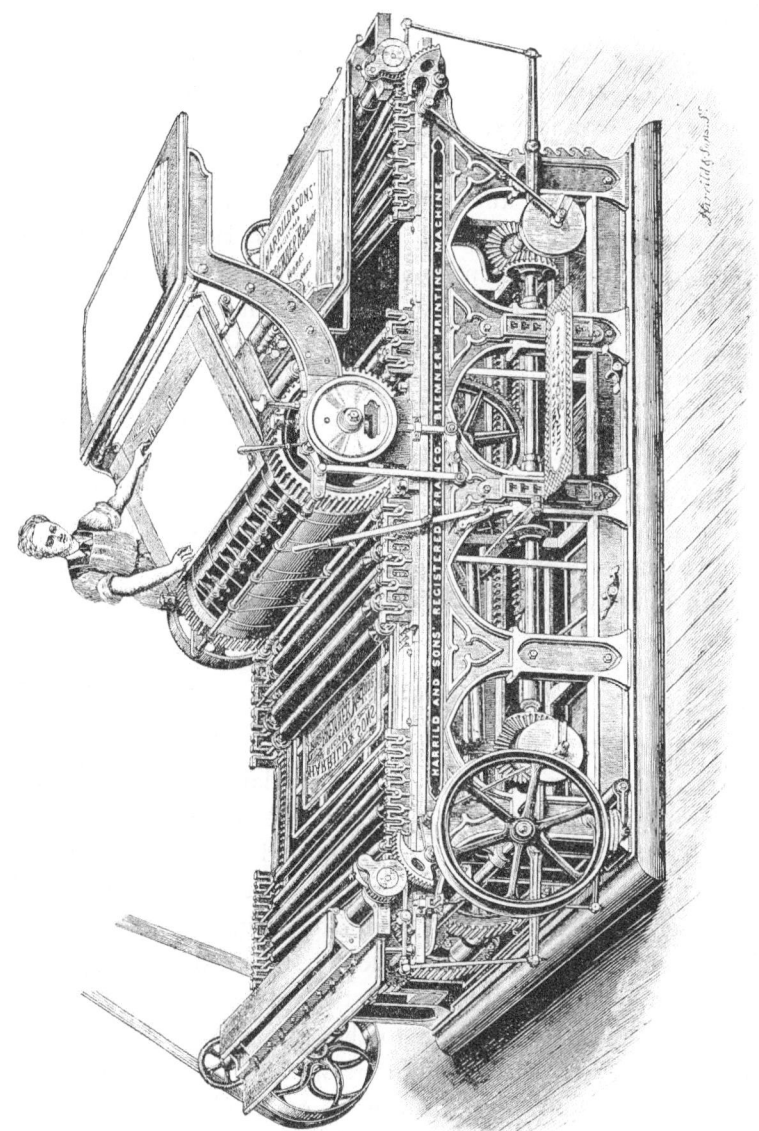

FIG 20.—THE FRANCO-BREMNER MACHINE.

the taking-off board underneath—above the supplementary inking table. A short series of tapes travel round the cylinder, and the front edge of the newly printed sheet is thus conducted from the grippers over tape-bars on to the heap. The lad has simply to arrange the sheet, as it is really automatically delivered.

As many as eight inkers, together with riders, entirely cover the forme, and inasmuch as the ink is supplied and distributed at two different points, the perfection of inking may certainly be attained—a small quantity of ink and a large amount of rolling.

In general construction it may be said to be identical with the ordinary Bremner. As will be seen from Fig. 20, it is substantially made and excellently fitted. It is slightly more expensive than the machine of the ordinary make, in consequence of the extra fittings, &c., required in connection with the cylinder and taking-off arrangements.

THE "*GRAPHIC*" MACHINE.

Designed by Mr. Parsons, and originally made by Messrs. Davis, especially to print the large and heavy cut formes of the *Graphic.* Prior to the introduction of this machine, one of the admitted defects of the Wharfedale was weakness, especially at the point of impression. To overcome this, the printing manager of the *Graphic* conceived the idea firstly of providing a sound iron bed-plate, upon which the whole of the frames could be bolted. In addition to this, the girder under the cylinder, together with the runners, were all materially strengthened—the whole insuring a positively unyielding and firm impression. The cylinder is also made proportionately heavier, and firmly secured in substantial bearings. The cylinder bearers are made of iron, so that

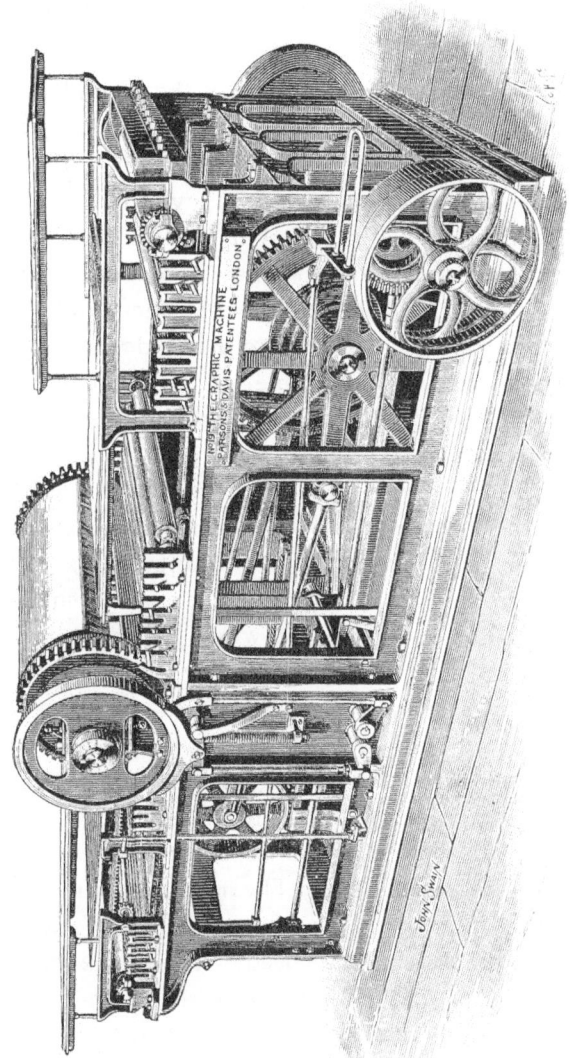

Fig. 21.—THE "GRAPHIC" MACHINE.

packing is practically impossible—in fact, not needed, as it is urged by the constructor that the cylinder should press flush to its bearers. It is claimed that, in consequence of the great strength of the cylinder and the supporting brackets, no "give" is possible; and as the cylinder is worked with hard packing, there is no inclination to "dip," even if the cylinder is supported above its bearers. In fact, if it is desired to work an extra-large forme, the bearers may be removed altogether. We are, however, of the opinion that the cylinder should be provided with support on either side, in addition to the journals, and it is desirable to allow the ends to run firmly upon the iron slips in the ordinary way, as there is then no probability of undue strain.

On the "*Graphic*" the grippers are placed slightly lower in the cylinder than on other one-sided machines. This we do not consider an advantage, as the space between grippers and board is necessarily smaller, and at times causes some trouble when the paper is in bad condition.

There are several minor features peculiar to this machine, but it may be practically considered to be generally identical with the typical one described. It will be noticed, however, on reference to Fig. 21, that the driving shaft is placed considerably lower than usual, and that the fly-wheel is small and solid. The inventor claims "strength and simplicity," together with capability of doing first-class cut-work at a good speed. Varied experience has justified this claim. Compared with other machines, it cannot be considered perhaps so elaborately finished, or so ornamental in appearance, but it must be remembered that polish and elaboration of design are not essential to good printing.

The "*Graphic*" is also made with a movable table,

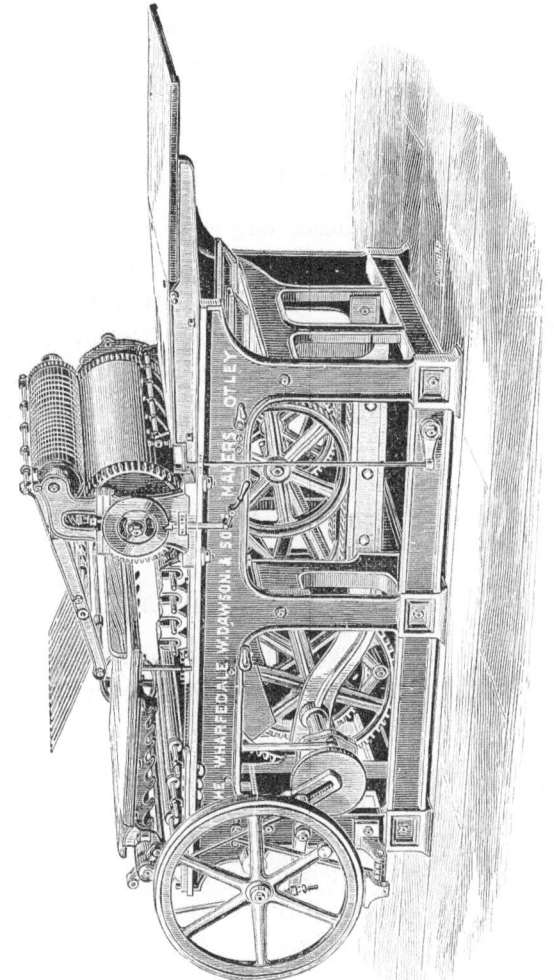

FIG. 22.—DAWSON'S WHARFEDALE MACHINE.

enabling it to be used either for type or litho printing. We must confess, however, that we do not admire "combination" tools, and would advise that when a machine of this make is required it be distinctly ordered for letterpress.

DAWSON'S WHARFEDALE.

One of the most deservedly successful and popular machines of this class. Every modern improvement is embodied, and being well balanced and fitted, it is eminently adapted for all descriptions of work. Some of the best book-work is produced on this machine, and the fact of Messrs Dawson being among the first makers of the stopping cylinder—to which they added the flyer taking-off apparatus—places their Wharfedale in a well-earned front position in the printing business.

PAYNE'S WHARFEDALE.

One of the original Wharfedale machines, which continues to maintain a foremost position in the trade. Like the Dawson, the Payne is fitted with all the appliances mentioned in the foregoing general description. It is light in construction, and in consequence requires comparatively small power to drive it. At the same time, it is judiciously strengthened at the points necessary to insure its being able to do heavy work. The inking arrangements are excellent.

THE QUADRANT MACHINE.

Originally introduced as a small jobbing machine, built on the Wharfedale plan, the demand justified the makers' (Messrs. Powell and Son) offering them in larger sizes, and

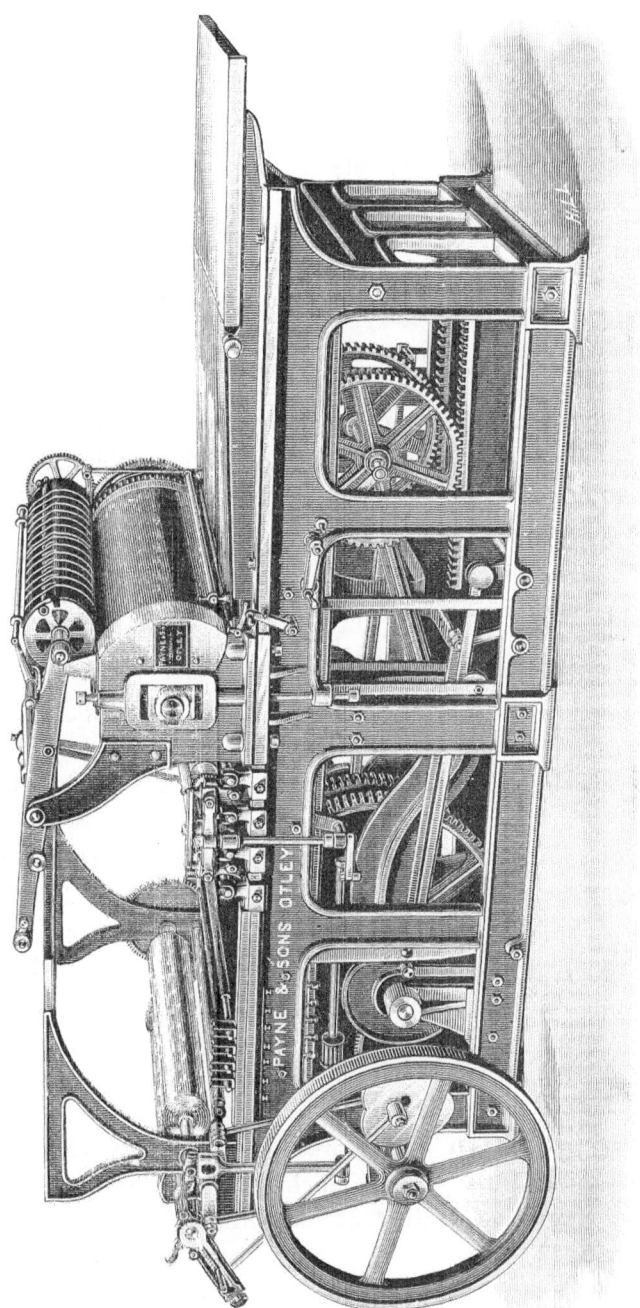

FIG. 23.—PAYNE'S WHARFEDALE MACHINE.

a quadruple demy has lately been made and has proved a well-deserved success. The first constructed were intentionally light, to enable them to be worked at a high speed, but the larger ones have been judiciously strengthened, and bear favourable comparison with others of the same class.

What may be termed the full-sized Quadrants are fitted with double inking motion, and have also a peculiar taking-off apparatus, consisting of a new cylindrical flyer, which delivers the printed side up without the assistance of tapes. It is also provided with double gearing, and additional supports under the cylinder. One of the advantages claimed by the makers is that it is a thoroughly reliable machine at a relatively low cost.

THE INGLE MACHINE.

An excellent and well-known jobbing machine, singular, perhaps, for the extreme simplicity of its construction. It is very light in its running, and probably takes less power to drive than any other machine of the same size. For light formes and general work it is eminently suited, and may be run at a high rate of speed with perfect safety. A large number are in use, and it may be found in the majority of commercial houses.

MESSRS. NEWSUM'S ANGLO-AMERICAN LETTERPRESS MACHINE.

This well-known Leeds firm was originally identified chiefly with lithographic machinery, and subsequently adapted the chief feature of the litho to the letterpress machine.

As will be seen from the illustration, this machine

FIG. 24.—NEWSUM'S ANGLO-AMERICAN MACHINE.

suggests almost in every respect the litho—having the overhead laying-on board, and the taking off board immediately underneath. As we have before mentioned, this principle is not entirely new, having first been introduced into this country by Messrs. Dutartre, of Paris. It is, however, more massive in its construction than its

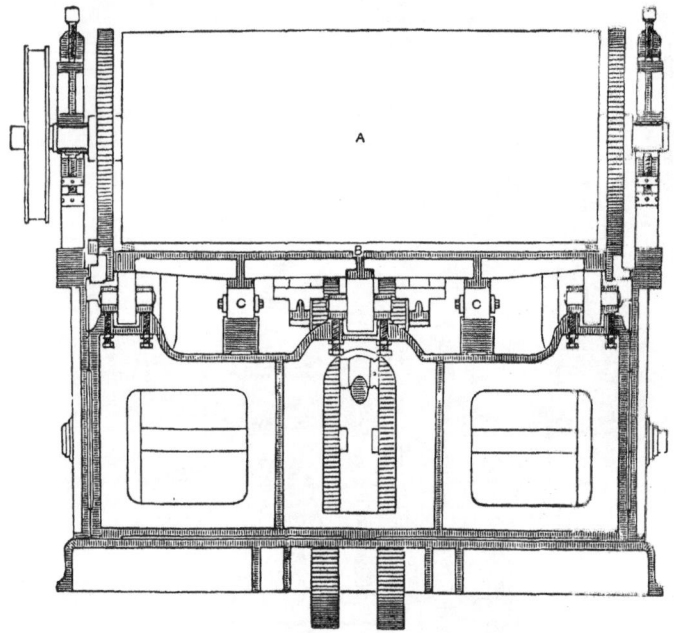

Fig. 25.—Cross Section of Newsum's Anglo-American Machine.
A Cylinder ; B Table ; C Impression pulleys and runners.

predecessors. But the whole is so well fitted and balanced that it may be worked at a high rate of speed on the best class of cut-work.

At least six inkers may be made to cover a full-sized forme, and arrangements are made for the accommodation of a series of riders.

This machine is to be recommended for high-class

magazine work, on which it is employed with a large amount of success. It may be anticipated that they will be found in all offices where fine-art printing is a speciality.

To afford some idea of its strength and solidity, we give an illustration of a cross section.

In addition to the foregoing machines, there are several which may be said to be in every respect excellent, and to possess great merit. The fact is, that an inferior Wharfedale machine, however cheap, would be an immediate failure, and therefore makers would not risk the initial expense of making patterns, unless they were thoroughly convinced that their machine would in some way compare with those of old and established reputation. Of course some makes are in special respects superior to others. This must be self-evident. For general purposes, the following machines may, with the greatest confidence, be relied upon. But inasmuch as they all may be said to be generally similar in construction, they call for no special remark. The makers, however, in each case are engineers of eminence, and can boast of having machines successfully running in almost every printing centre :—

FURNIVAL'S IMPROVED WHARFEDALE,

manufactured at their well-known works at Reddish, near Stockport, is well and substantially built. The table is supported by four rails or girders, and special attention is given to the impression pulleys, so that any reasonable amount of impression may be used. A cross-stay or girder, especially strong, is also fixed under the centre of the cylinder. The cylinder and grippers of this machine are so arranged that the sheet may be laid to the mark

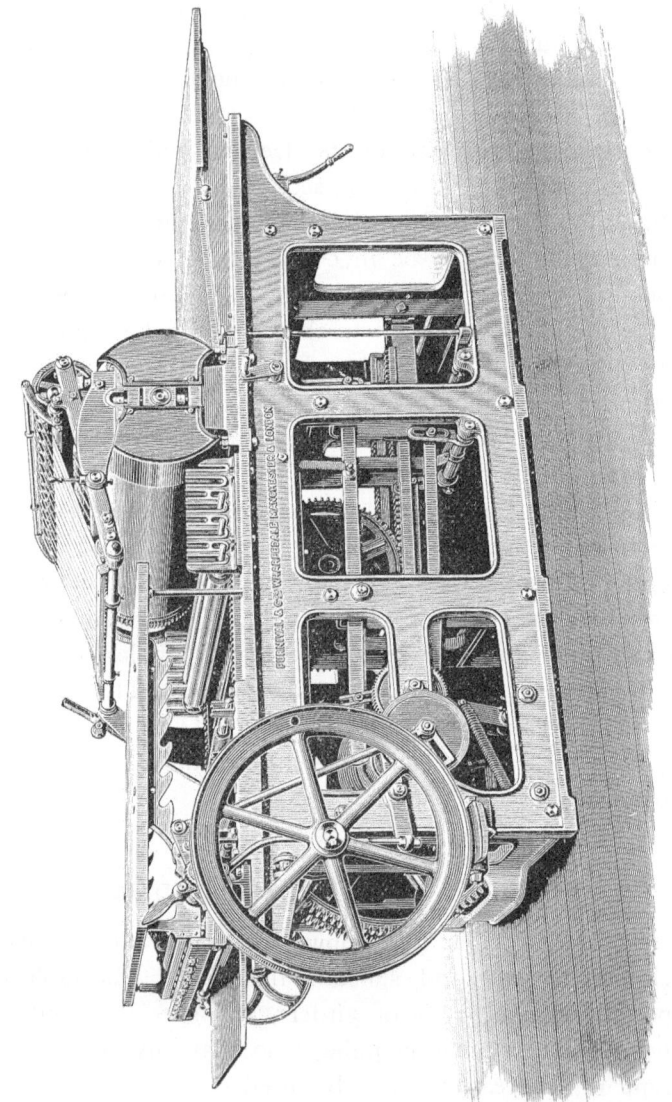

FIG. 26.—FURNIVAL'S WHARFEDALE.

FIG 27.—THE RELIANCE WHARFEDALE.

immediately the preceding one has been taken, allowing the boy ample time.

THE LEADER machine, manufactured by the Birmingham Machinists Company is singular inasmuch as the large wheel and connecting-rod are dispensed with, and the tables are driven by a *horizontal driving disc* placed immediately underneath. Some years ago several machines were constructed on this principle, but were not very successful, owing to the " play " which occurred after some little wear. The makers of the Leader have designed and patented a guide, fitted with double action anti-friction runners. By this means any side movement is prevented, and the whole being adjustable, the original defect in this horizontal motion has been successfully overcome.

The gearing, which is singularly compact and simple, lies immediately under the inking table, and it is claimed that as it is fixed to the body of the machine (which, in the case of small sizes, is made in one casting) no vibration or jar is possible.

The Leader is well adapted for the Colonies, as it is easily packed, and may be re-erected quickly, and without the assistance of fitters, &c. The ordinary flyer is used for taking off.

THE RELIANCE, by Fieldhouse, Elliott and Co, of Otley.

THE CLIMAX, by George Mann and Co., of Leeds.

THE CAMBRIAN, by Pullan, Tuke and Co., Leeds.

SALMON'S, of Manchester.

SEGGIE'S NEW " EDINBURGH."

CUTHBERTSON'S PATENT COUNTER

is an ingenious apparatus, which may be easily fitted to the cylinder of the Wharfedale by means of a stud or a

bracket. In addition to dials indicating the number
of impressions, separate ones are also supplied, that the
exact hour and date may be registered when the job
was commenced, and thus the time occupied and the

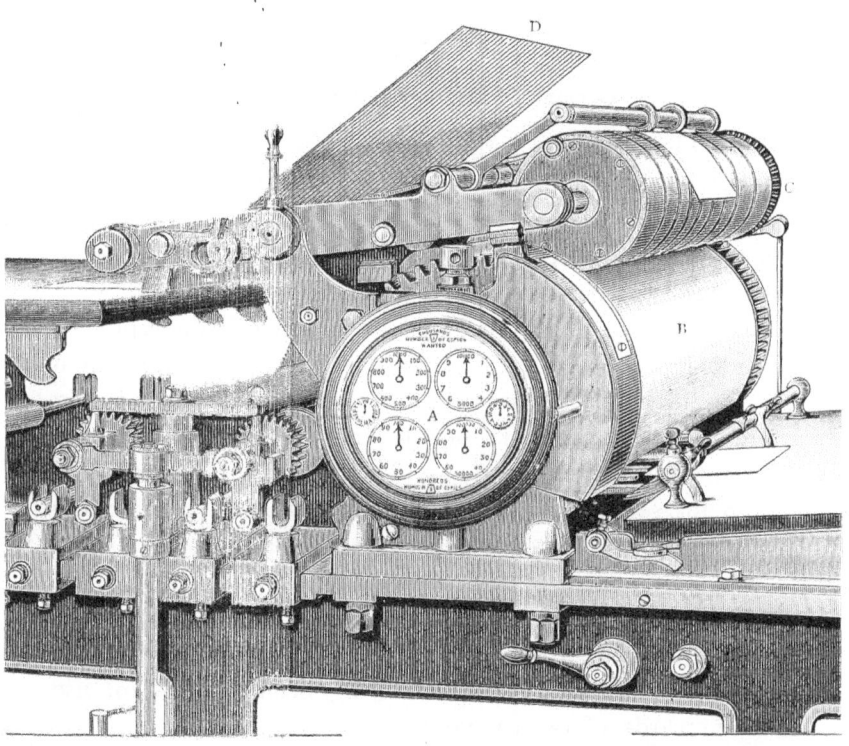

CUTHBERTSON'S PATENT COUNTER.

A Patent counter; B Impression cylinder; C Taking-off drum; D Flyers.

number printed may be calculated at a glance. When
making ready or running waste, it may easily be thrown
out of gear. Each counter is carefully tested up to
100,000. Although, of course, not indispensable, they
may be considered a useful addition to the machine,
especially when long numbers are printed.

CHAPTER IV.

THE TUMBLER OR MAIN MACHINE.

General Description and Peculiarities.

THE Tumbler or Main was a really reliable and successful machine some considerable time anterior to the introduction of the stopping cylinder, and in fact may be regarded as having afforded many practical suggestions upon which the latter was constructed.

The Tumbler was invented by Thomas Main, a printer, who made several machines; but he suffered from severe financial difficulties, and had to encounter strong prejudice on the part of the trade. Subsequently the patent was taken over by Mr. Conisbee, an engineer, who for some long period possessed almost a monopoly in the making of this class of machine. It should be stated, in justice to his work, that many are at present running, and, while perhaps not suitable for really first-class printing, are well adapted for good commercial work, notwithstanding their having been in constant use for upwards of twenty years.

The cylinder of the Main machine "tumbles," or reverses, after the impression is taken, and returns with the tables. This is effected by cog-wheels on each end of the cylinder working in gear with the corresponding racks on either side of the table. Both the cylinder cog-wheels are keyed to the shaft. In order to admit of the forme clearing the blanket on the return of the table, he cylinder is lifted up slightly—being immediately depressed on to its bearings when the grippers have grasped the sheet for the succeeding impression. This is accomplished by two steel rods attached to bell-cranks on the outside of the cylinder shaft—one on each side. The motion is conveyed by a rod from the cam shaft, which works another shaft at right angles, running from side to side of the machine under the cylinder, and resting upon the bottom girders of the side-frames. The cylinder-rods, being attached at the other extremity to the cranks on the lower shaft, force the cylinder up and down, being regulated by the action of the driving cam.

Generally speaking, the construction of this machine excepting the cylinder motion, is somewhat similar to that of the ordinary Wharfedale. The tables are driven from the main shaft, on which are placed the various cams, which regulate the points, movement of laying-on board, double inking, &c. This latter improvement was added, together with adjustable points, &c., after the death of the inventor. The construction of the laying-on board was originally by no means so perfect as in those fitted to the later machines. The front part was so made as to move backwards and forwards. Immediately the sheet was taken by the grippers the whole of the front portion of the board moved back some little distance The fresh sheet was then laid in position, flush to the front marks, and when the cylinder returned, a forward motion was imparted—placing the sheet within grasp of the grippers.

Although, as before stated, the Main enjoyed considerable popularity prior to the introduction of the stopping cylinder, it necessarily gave way to the improved machine. The advantages possessed by the latter were plainly apparent. The wear and tear of the Tumbler was necessarily greater. With the Wharfedale we have a cylinder simply revolving when the actual work is being done, but in the case of the Tumbler, the cylinder performs two revolutions to every impression —once to give the impression, and again to admit of the forme returning preparatory to taking the next sheet. The strain exerted is also greater, as the cylinder is brought to a dead standstill, and immediately carried back by the tables ; whereas the Wharfedale cylinder is eased by the automatic brake, just before it has completed its revolution.

As the cylinder is really supported by the stout steel rods on either side, the multiplication of bearings is apt to give trouble, as the brasses will sometimes wear unevenly, causing a slur on the entry. This defect may be to some extent remedied by judicious packing of the cylinder bearers. But when the bearings of the shaft are much worn, the blanket or cylinder sheets will become blackened by the forme, which is, of course, owing to the cylinder not being raised sufficiently high to admit of the forme returning without slight contact. When this is the case, it is far better to have the bearings readjusted by an engineer, than to attempt to obviate the defect by multiplying the cylinder sheets.

Very much less time is allowed for the removal of the sheet than on other machines, as, practically, the cylinder is never stationary—returning immediately the impression is complete. Directly the extremity of the sheet appears above the taking-off board, it must be held and carried over quickly on the heap, otherwise it is apt to be taken round the rollers.

With the Scandinavian and Platen, we may perhaps now relegate the Main to the past—not because it is a faulty machine, and incapable of good service, but because others of more modern and improved make have gradually and surely superseded it.

CHAPTER V.

THE TWO-COLOUR MACHINE.

General Description of the Wharfedale, Bremner, and Payne's Two-
Colour Machines—Huber and Hodgman's Two-Colour Machine.

THE two-colour machine is in general construction
practically identical with the ordinary Wharfedale. It is,
however, provided with two coffins with ink tables and
inking apparatus at either end, as in the case of the
perfecting machines. The cylinder is fixed across the
centre of the side-frames, and works in gear with the
table racks on each side. The distance from the edge
of the ink table to the partition dividing the coffins is
equal to the circumference of the cylinder, which thus
performs its first revolution when this point is reached,
and, continuing with the table racks, completes two
revolutions, and stops when the tables arrive at the extremity.

The machine occupies much more space than the
Wharfedale, as the frames have to be considerably
lengthened to allow for the extra tables, &c. The
gearing and various movements are the same as described in the previous chapter. Five inkers cover the
entire forme at either end, so that this machine is
eminently adapted for first-class work.

The tables are usually made of marble, which is
better suited for coloured inks than iron. It is advisable to have a row of gas-jets suspended beneath the
slabs at either end, as by the application of a gentle
heat peculiar inks may be kept in better condition, and
more evenly distributed in cold weather. The taking-off

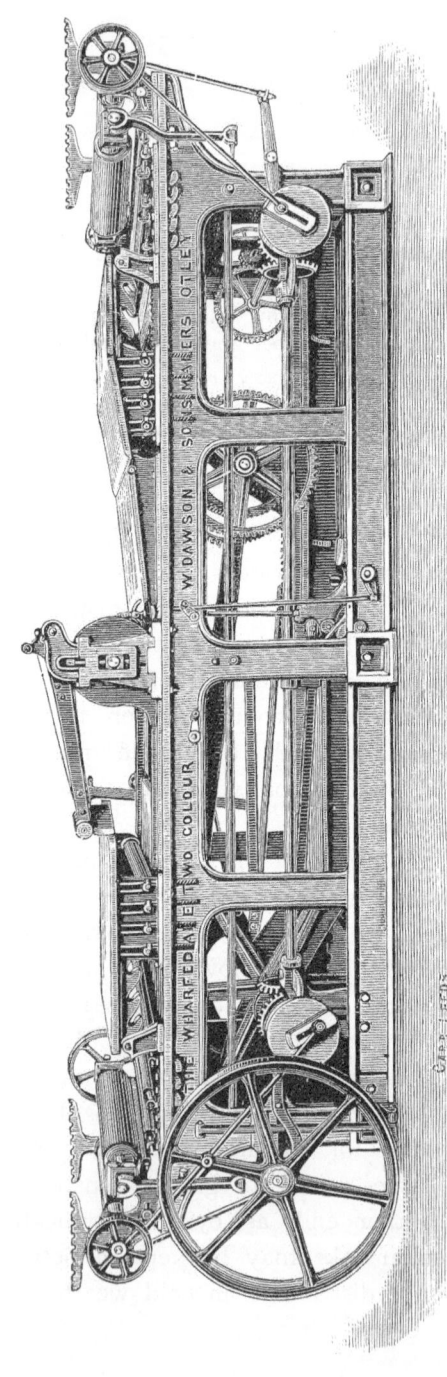

FIG. 29.—TWO-COLOUR MACHINE, WITH TAKING-OFF APPARATUS.

apparatus may be fitted ; but perhaps this is not advisable, as when heavy coloured work is being done, it is safer to employ a boy to carefully lift the sheet than to risk a rub on the tapes, which is sometimes unavoidable when there is a large surface covered with colour.

Supposing the formes to have been accurately made up and properly laid, the register is necessarily perfect, inasmuch as the grippers retain hold of the sheet during the printing of the two formes. The sheet is fed on to the cylinder, which revolves twice before the grippers open again. For two-coloured titles, bills, broadsides, and similar jobs, it is well suited ; but its economy is to be questioned unless a continual supply of such work be forthcoming, as the actual time occupied with the printing is nearly double that of a single machine, the travel being nearly equal to twice that of the latter. Four hundred perfect impressions per hour may be considered a fair and satisfactory speed. This must necessarily be regulated, however, by the body of colour required, as in heavy broadside work it would be impossible to lift even at this speed—as although the number mentioned is certainly moderate, the tables and cylinder are travelling equal to the rate of 800 on the ordinary machine.

It is doubtful if it pays to put a job of less than 5,000 copies upon the two-colour. Two formes have to be made ready, and it takes just the same time to adjust the ductors, rollers, and ink as it would if two machines were employed. They are usually put down for peculiar work having long runs, and may be considered under these circumstances a profitable investment.

The two-colour machine is made by Messrs. Dawson and Sons, Messrs. Harrild and Sons, and Messrs. Payne. They may differ in a few minor details, but all may

Fig. 30.—HARRILD'S TWO-COLOUR BREMNER MACHINE

be relied upon, as being able to produce really good work.

Messrs. Huber and Hodgman, of Taunton, U.S.A., have recently introduced a two-colour machine constructed upon entirely new principles. We take the following description of the same from an American journal.

" The general appearance and principles are the same as any modern two-revolution cylinder press. The cylinder has two impression services instead of one, and the bed carries two formes of type, the ink being supplied in the usual manner by two fountains located at each end of the press. Above and behind the cylinder is placed a series of wheels, called transfer wheels. The operations of the press are as follows:—The sheet is fed in the usual manner from the feed-board on to the first impression surface of the cylinder, and passing around, takes an impression from the first forme of type, after which it is taken by the gripper on the transfer wheels, and after making one revolution is delivered back again to the cylinder on to the second impression surface, where it receives its last impression from the second forme of type, and is delivered, printed-side up, on to the table in front of the cylinder. In the meantime another sheet is fed on to the first impression surface, so that the press delivers a sheet with two colours every two revolutions of the cylinder—in other words, accomplishes twice the work of an ordinary machine in the same time. Although the press is especially designed for colour work, it is adapted to do all kinds of first-class printing. The press will print a sheet 48 by 32 inches; is 18 feet long over all, and requires about one-horse power to run it."

J

CHAPTER VI.

THE TWO-FEEDER SINGLE-CYLINDER MACHINE.

Dawson's—The " *Graphic* "—Bremner's—and Payne's Two-feeder
Machines.

SOME time prior to the introduction of the Dawson
two-feeder Wharfedale, Messrs. Middleton made a two-
feeder machine which was very successful. Although
it printed on one side only, it was provided with two
small cylinders. These cylinders rose and fell alternately (somewhat after the style of the Anglo-French),
to allow the forme to clear one while the sheet was
being printed by the other. This motion was effected by
simple cams at the base of the machine, working in the
frames supporting the cylinders. The coffin and tables
were driven by an upright spindle and rack. Although
many are at work at present in small newspaper offices,
and doing good work, general preference is now shown
for the improved single-cylinder two-feeder, originally
the invention of Mr. Dawson, of Otley.

It is obvious that a two-feeder machine must print
both ways—as the forme is travelling out and returning—therefore the " pitch " must be exactly the same
at both ends. In introducing their new machine, Messrs.
Dawson described it as being the "only Wharfedale single
cylinder two-feeder gripper machine yet known that will
print any and every size newspaper below full size either
as a one- or two-feeder."

The advantage of being able to work a newspaper
sheet of any size into the same two-feeder machine

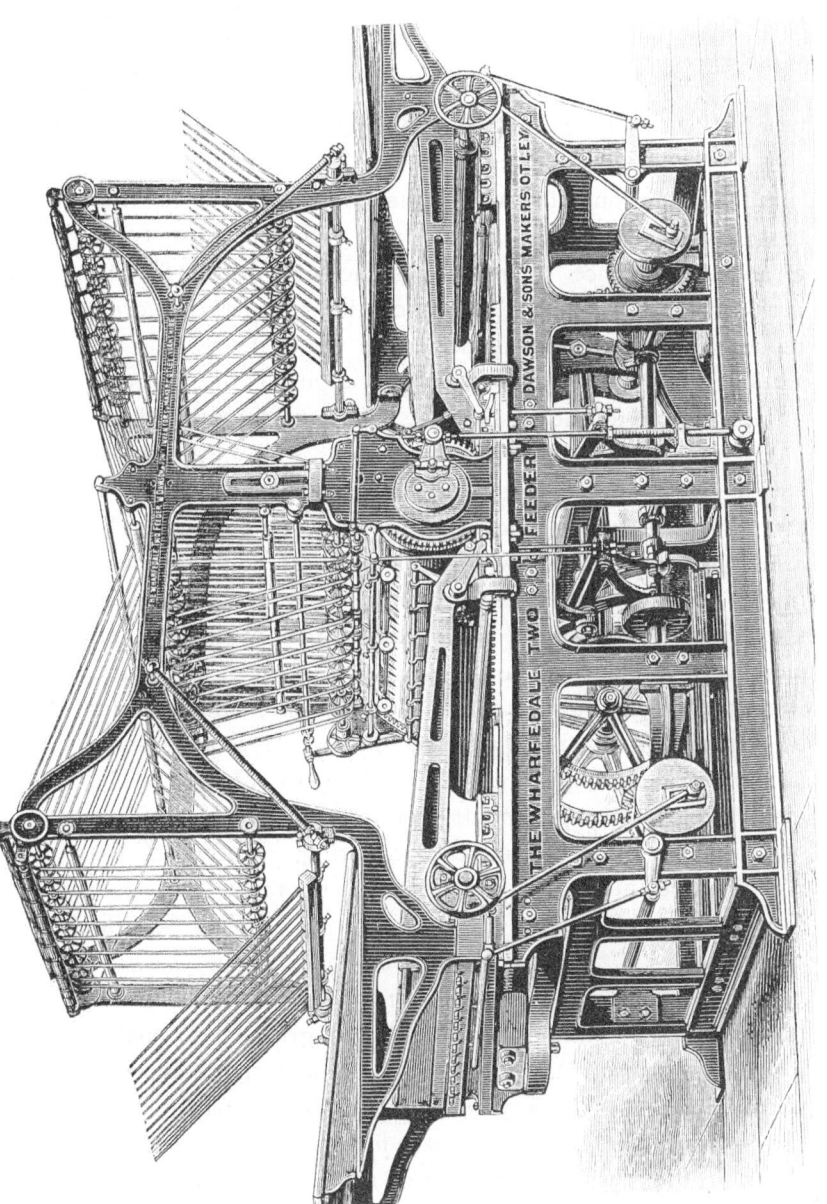

FIG. 31.—DAWSON'S WHARFEDALE TWO-FEEDER.

J 2

(especially in country offices, printing journals of different styles), was early perceived in the trade. The problem practically resolved itself into the change of position of the grippers. As the cylinder in these machines has a tumbling motion similar to the Main, it is clear that in the case of a large forme the cylinder must travel a greater distance to present sufficient surface for impression, and *vice versâ.* Hence the grippers, in order to come into position to seize the next sheet, must be set further from or closer to the edge of the cylinder, in proportion to the size of the forme to be printed.

Before this machine was invented, the only way in which the difficulty had been obviated was by constructing the cylinder of loose pieces, or segments, by removing and shifting which the grippers could be fixed in the required position, although of course only fixed sizes could be printed. The difficulties attending this clumsy arrangement, and its great want of accuracy, were perceived, but accepted as inevitable, and a series of experiments were made by Messrs. Dawson, some account of which appeared in the *Printer's Register.* The result of these experiments was the discarding of the old principle entirely, and the patenting, in 1874, of the machine we are about to describe, or rather to allow Messrs. Dawson to do so in the terms of their specification.

"The object of our invention," say the makers, "is to construct a cylinder furnished with grippers to enable a machine of the class named (Wharfedale single-cylinder two-feeder) to print any and every size of paper, from the smallest to the largest size, for which it is made as a two-feeder, without the use of movable pieces, or segments, for the printing surface.

"We accomplish this in the following manner:—The

surface of the cylinder is made rather larger than the
largest sheet to be printed. The gripper shafts are
placed *inside* the cylinder (instead of outside, as heretofore), and are supported at the ends by suitable brackets
fixed in radial slots on the ends of the cylinders, or
on the outside of cylinder wheels, or inside of rim of
cylinder.

"Instead of the cylinder being furnished with a number of small grippers as heretofore, it is fitted with four
thin continuous grippers of suitable shape and breadth,
each extending from the centre margin tape on the
cylinder to the outside of the largest sheet to be printed ;
on each of these grippers are fixed or cast projecting
bosses, one at each end, which are bored and furnished
with set-screws to secure them to the gripper shafts.
Slot holes are made round the circle of the cylinder
(through which these bosses are fixed on the shaft
inside) of sufficient length to allow of the shafts being
moved inside while the grippers are moving outside over
the surface of the cylinder to any point required."

The plan of construction of the cylinder will be
easily understood by reference to the figures, which represent respectively an end view of the cylinder (Fig. 32)
and a plan of the same (Fig 33) showing the application
of the system. The lettering is identical in both, the
same parts in plan and end view being indicated by
the same letters. Thus A A in both drawings are the
cylinder wheels ; B B the brackets which support the
gripper shafts D D, which are placed, as already described,
inside the cylinder. Round the surface of the cylinder
and parallel with its cross-section are cut the slots
M' M', in which the bosses E E, fixed upon the gripper
shafts, travel backwards and forwards, carrying the
thin plate grippers F F. The slots are enlarged at

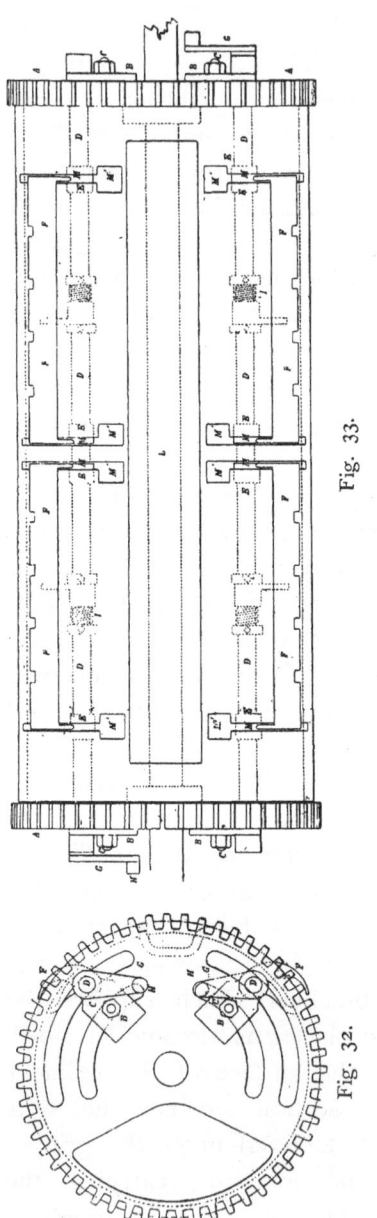

Fig. 33.

Fig. 32.

one end for the purpose of allowing the bosses to pass through in fixing the gripper shaft in position. Thus the whole shaft, with its bosses and plate grippers, can be readily shifted backwards and forwards along the length of the slot, and the grippers set at whatever distance apart the size of the sheet to be printed may require. The grippers are opened and shut by means of a cam or tappet, which, as the machine works, operates upon the small bowls or runners H H, which terminate the levers G G, fixed upon the ends of the cylinder shafts. I I are spiral springs fixed upon the shafts, the action of which closes the grippers when the cam ceases to operate upon the bowls H H. This, of course, takes place when the paper is fed into position under the open gripper.

When it is wished to alter the position of the grippers upon the cylinder the bolts C C, which secure the brackets in place on the end of the cylinder, and these latter can then be moved, carrying with them the gripper shaft, the bosses sliding in the slots upon the surface of the cylinder, and the gripper plate moving outside until the required position is reached, when the bolts are again tightened and the brackets secured in their new place. To facilitate the setting of the grippers, most machines have an index or scale upon the surface of the cylinder, so that the setting can be done for any given size of paper by reference to it.

The length of travel of the cylinder is of course regulated by the table racks in which it is in gear. The table is driven to and fro by a crank from the large wheel geared to the driving shaft in the ordinary way. The end of this crank is secured in a slot in the wheel, and when it is necessary to alter the pitch to accommodate a certain sized forme, its position is shifted, thus shortening or lengthening the stroke as desired. That this may be done with accuracy, a series of marks are cut on the side of the slot, which correspond exactly with identical indications on the sides of the gripper openings on the cylinder. By this means an alteration may be both accurately and speedily made.

The machine may be readjusted easily in half an hour.

The popularity of the Dawson two-feeder Wharfedale has been very great. We believe that hundreds are at present in work throughout the country. They may be run safely at a speed of 1,500 each board.

The "*Graphic*" two-feeder machine has also achieved a large amount of success. It differs in many respects from the Dawson, and possesses several distinctive features.

Like the "*Graphic*" single-cylinder, it was built specially for printing cut-work. It is very solid in construction, the bed-plate being adopted, and a heavy girder placed beneath the cylinder to insure solid impression. An extra series of runners is also placed under the centre of the table. These machines have a longer travel than the Dawson, and consequently the cylinder is larger. This insures a larger number of inkers passing over the forme.

The sheet is fed into the grippers much higher up the cylinder than in the case of the Dawson. Another peculiarity is that the sheets are delivered to the laying-on board, face up, on the same side as fed-in. In other machines the sheet is delivered on the opposite side.

The "*Graphic*" two-feeder is only made to print a certain sized sheet, the grippers being constructed and arranged in a similar manner to the ordinary single cylinder. If, however, it is desired to print a smaller forme, it may of course be done by working the full-sized sheet and cutting to waste.

This machine can be used as a single-feed stop cylinder, the alteration being speedily effected by means of cams on the main shaft.

It may be mentioned that the cut-formes of the *Graphic* are worked on this two-feeder—thus demonstrating its capabilities.

Messrs. Harrild make a two-feeder machine—the Bremner. It is largely used by provincial newspapers, and is also employed by the *Illustrated London News.* As is characteristic of all Messrs. Harrild's work, it is in every respect excellent. With cut steel racks and substantial support to the cylinder and table, it is to be commended for any class of printing in which speed is necessary—either newspaper or illustrated work. A large number have been made, and are justly considered to be powerful rivals to those already mentioned.

Messrs. Payne, also, have a two-feeder Wharfedale, which, while possessing few peculiarities, is also extensively used by London and provincial printers.

It may be said that two-feeder machines are generally erected for special work. They are adapted for newspapers of comparatively limited circulation. The advertisement formes are usually made up some hours before the inner pages containing the news, and may be thus worked off before the latter are ready to follow on the same machine.

CHAPTER VII.

THE DOUBLE PLATEN.

General Description of Hopkinson and Cope's Double Platen—The Napier Improved Double Platen.

THIS machine is really a steam press, the pressure being imparted by a platen in a similar manner to the hand-press. Before the improved Wharfedale was introduced, nearly the whole of the best class of work was printed on this machine, and a strong prejudice existed against the adoption of the cylindrical mode of impression, as it was professed, with some degree of truth, that it was impossible to obtain sharp, clear-cut work by means of the newly introduced cylinder. This was owing to the fact that the relationship between the cylinder and the tables had not been properly decided, and perhaps the workmen were not so expert in handling the cylinder as the platen. However, after some years it was demonstrated that better work, owing to improved rolling, &c., could be produced on other machines than the platen, and the latter may now be considered to be virtually superseded. But as many are still employed, we think that a work on Printing Machinery would be incomplete without some reference to their general construction.

The platen itself is placed across the centre of the side-frames. It is supported by a powerful beam, one end of which is held firm in a bearing, while the other works freely up and down between guides. Although the travel of the beam is irregular—*i.e.*, one end moving

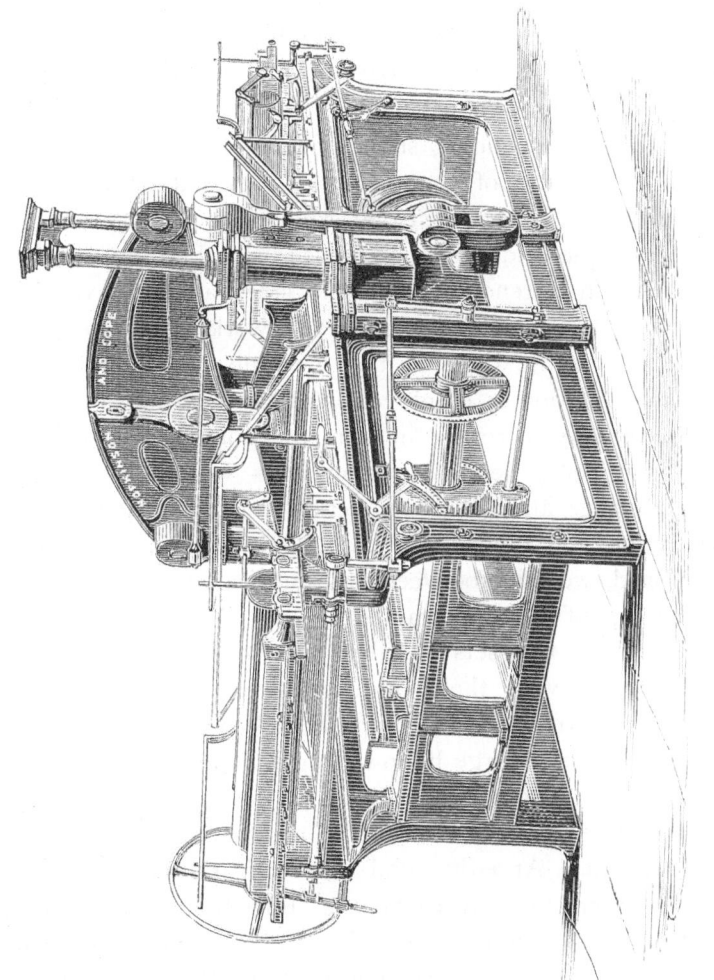

FIG. 34.—HOPKINSON AND COPE'S DOUBLE PLATEN.

about twelve inches up and down, while the other extremity
is held stationary—the platen beneath, supported in the
centre by strong bearings, maintains a perfectly even
motion up and down. This is owing to its being held
firmly between V-shaped grooves on either side, into
which fit corresponding shapes fixed on each end of the
platen. The top of the platen is provided with a cup,
into which fits flush a strong rounded bolt attached to
the under-side and centre of the beam. The end of the
latter, which extends outside the side-frame, derives its
motion from a powerful steel crank working on the end
of a shaft at the base of the machine. Although the
bell-crank motion is employed, and the crank at its base
has necessarily a rotary motion, the portion attached to
the beam being held by strong bolts working freely in
their bearings allows of the action to the platen being
strictly up and down, the side of the beam, as before
mentioned, working between well-lubricated guides fitted
on the top of the side-frames.

There are two sets of tables, one at either end, both,
however, receiving their motive power from the same
mechanical arrangement. Supported beneath one end of
the machine is a large hollow drum, about twenty inches
in diameter. Around this drum is a helical groove, about
one inch deep, arranged similar to a right- and left-handed
screw-thread. At either end of the drum the worm, or
groove, extends evenly round the entire body. A shuttle,
or steel shape, slightly pointed at either end, travels in
the groove, and the rotary movement of the drum moves
the switch forward or backward until one extremity is
reached, when, running into the perfect circle, it is at
rest for a period—until, in fact, the drum has performed
one revolution, when by a "shape" it is again diverted
into the worm, and travels to the other end of the drum,

where a rest again occurs. (See Figs. 35 and 36.) Above the grooved drum is a flat, square slide, supported between two steel guides. The switch before referred to is attached by a bolt to the under surface of

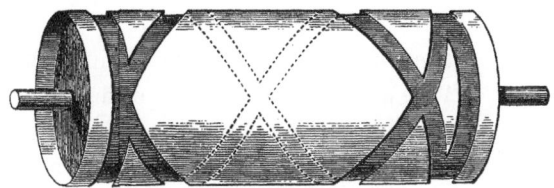

Fig. 35.—Helical Drum.

the slide, so that the latter has a steady backward and forward motion, being held firmly between the guides. In the centre of the slide is a hole, into which drops a steel bolt fastened to the under-side of the table.

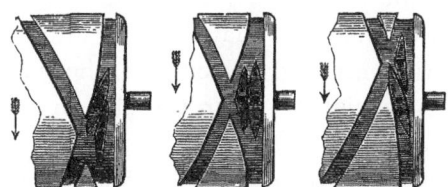

Fig. 36.—Showing the various Positions of Shuttles.

Thus, when the bolt is allowed to drop into the slot on the travelling slide, the table is moved towards the platen, the coffin resting while the switch is travelling round this circle, enabling the impression to be taken. Immediately the platen rises, the switch, directed again along the worm, carries the table to the other extremity, when it becomes again stationary for a short time, to admit of the sheet being lifted and another placed in position.

Attached to one end of the movable slide is a steel

shaft, extending under the impression bed to a similar slide at the other end of the machine. The second slide is supported by and travels along guides under the coffin in the same manner as the one above the grooved drum. Thus the two slides derive their motion from the same mechanical arrangement, and when one is at rest at the extremity of the machine the other is in position near the platen. When the fixed bolts under each of the tables are allowed to drop into their respective slots in the slides, the coffins are carried backwards and forwards alternately under the platen. While one impression is being taken—the tables being at rest at the other end—the printed sheet at the other extremity is removed, and another laid upon the frisket. Thus the tables are alternately moved under the platen, and come to a dead standstill immediately before the latter descends to its lowest point.

The Platen machine may be put into motion without necessarily printing, as long as the table bolts are held sufficiently high to clear the slot in the slide. At either end a spring striker is provided, communicating with the bolt under the table. By loosening a self-acting spring, and raising the handle, the bolt drops sufficiently low to fit into the slot of the movable slide, when the table is carried under the platen and back again. Thus either or both ends may be worked as desired.

The frisket and tympan, each separate frames, are secured to the impression extremity of the coffin by gun-metal hinges or joints. These run upon steel bars extending from the side-frame near the bed of the platen to the end of the iron frame supporting the laying-on board. When the coffin is at rest at its extremity the tympan and frisket lie at an angle of about 45° on the bars. The tympan is thrown back in an almost perpendicular position against its support, the frisket remaining on the slide to admit of the fresh sheet being laid. Immediately the tables are about to return for the impression, the layer-on grasps a small handle on the tympan, and allows it to fall on the frisket, when both travel down under the platen, when the forme, frisket, and platen are necessarily on the same level. Although the forme moves back in a perfectly horizontal position, a "lip" on the fore edge of the frisket frame guides it and the tympan (which lies on top) up the slides, so that the sheet and forme part company immediately the printing is completed. The taker-off lifts the tympan on its return to allow time for the layer-on to fly his sheet on to the frisket.

The layers-on stand on the same side, so that one of them is compelled to adjust the sheet left-handed. One taker-off must, of course, also work left-handed.

The inking arrangements at each end consist of three inkers placed in their forks near the bed of the platen and entirely cover the forme—resting on the inking table when the impression is being taken. The inking drum, parallel with the ductor, is frequently adopted, to assist the distribution. On some machines small mouse rollers are employed, which travel backwards and forwards along the extent of the drum. When the ink is deposited by the second vibrator upon the table, it is well and evenly distributed.

The mode by which the whole of the various parts of the machine are put in motion really consists of a shaft driven off the main driving shaft by a small spur wheel. A bevel wheel near the centre works into a corresponding cog upon the shaft of the helical drum, while the extreme end, upon which is welded a powerful bell-crank (outside the side-frame), moves the beam up and down. The machine is therefore really worked from the

second shaft referred to, except the ductor and distributing drums, the former of which is driven by thin
rods extending from either a small bell-crank, or sometimes an eccentric on the main shaft inside the riggers,
and the latter by a gut band running round a groove
on the same shaft.

These machines were usually made to take a double-
demy forme, but some were constructed to take a
double-royal or four-crown sheet.   The large size, however, could not be considered so successful, as the
platen, being so large and heavy, was liable to become
unsteady after continuous work.   Added to this, the
enormous power required to impart sufficient impression, together with the extra work imposed upon the
boys in handling the tympan and sheets, militated considerably against the adoption of the larger make.

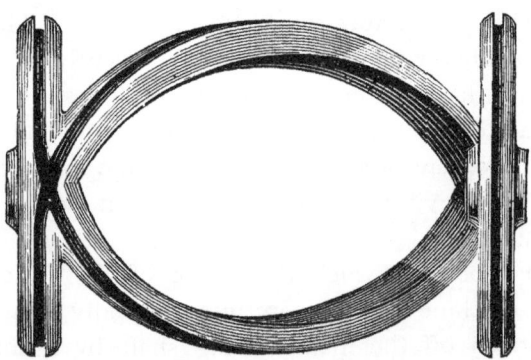

Fig. 37.—Helical Drum of Napier Platen.

THE NAPIER PLATEN.

The Napier Platen may be said to be far superior
in almost every respect to the one previously described.

K

Fig. 38.—THE NAPIER DOUBLE PLATEN.

The grooved drum is employed for the propulsion of the tables, but although, generally, the various movements are the same, the result is attained by altogether different mechanical contrivances. The beam is dispensed with, and the platen, consisting of a solid casting, is moved up and down by steel cranks worked by knuckle-joints on either side. The inking is excellent, the rollers passing over the surface of the forme four times between every impression.

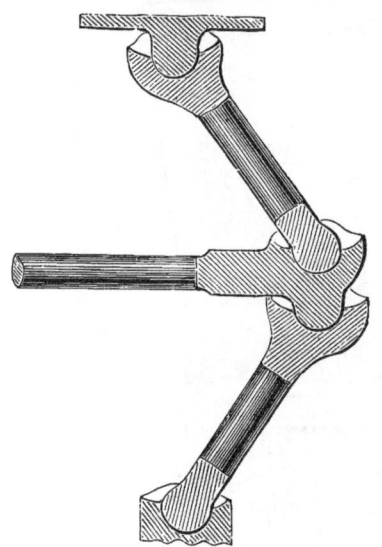

The main driving shaft is at the extremity of the machine, and by means of a bevel wheel imparts a motion to the spur wheel which works the grooved drum similarly to the ordinary platen. The platen itself is raised and depressed by a pair of connecting-rods, one on either side of the side-frame,

Fig. 39.—Knuckle-joint.

working a knuckle-joint attached by steel rods running through and secured to the head of the platen. The fixed point of the joint, as in the case of the Anglo-French, is at the top, and when the connecting-rod is at the extremity of the crank on the shaft, the platen is forced down upon the forme. This method of giving the impression is in every respect better than that employed on the ordinary platen. There is an absence of jar when the platen descends upon the surface of the forme, as the head is held rigidly on either side by identical rods, dispensing with the ingenious but somewhat complicated cross-beam and centre-bolt. Greater pressure may be also obtained with far less driving power (estimated at one-half) by the aid of the series of levers constituting the knuckle-joint. This may be easily tested by the workman, who can, with comparatively little effort, obtain a pull on the Napier by turning the fly-wheel, while in the case of the ordinary platen this would be absolutely impossible.

The platen is counterbalanced by a series of heavy weights, supported by gut bands. The em-ployment of the weights materially helps to balance and steady the platen head in its rise and fall. This

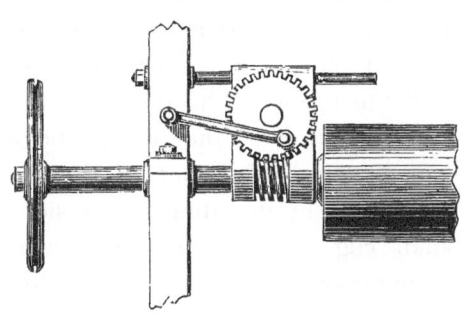

Fig. 40.--Vibrating Inking Drum.

is necessary on the Napier, as the whole of the head is moved entirely by the side rods, and lacks that partial support afforded to the beam of the ordinary platen.

Although the tables are moved by a slide travelling in the grooved drum, a clutch is employed, instead of the spring pin, working in the slot. This has been found to be a decided advantage over the original plan.

The inkers and wavers are placed in a movable frame on the top of side girders, and held on either side by broad steel bars, extending down and secured to the base of the inside of the side-frame by a bolt. The motion backwards and forwards is imparted by a steel bar running horizontally from a large cam on the driving shaft. A slot on the under-side of the bar fits on to

K 2

a bolt on the inside of the upright frame, and the whole inking apparatus is thus moved over the surface of the forme. When the end is struck off, by means of a small supplementary lever, the slot of the horizontal bar is lifted from the pin, and the inking frame becomes stationary. The cam is so made as to move the frame over the surface of the forme twice between every impression. Each end is provided with a separate horizontal bar with its cam, so that they work independently. The wavers on this machine are raised slightly above the level of the inkers, to prevent their coming into contact with the forme. The ductor is turned by a gut band.

Like some of the platens before referred to, a steel drum is placed in front of the ductor, to which the ink is supplied by its vibrator. A side motion imparted by a small cog-wheel moving in a worm on the shaft, tends to improve the distribution. No independent vibrator oscillates between the table and the under-side of the ink-drum, as the end waver in the ink-frame travels up a slightly inclined plane at the extremity, and takes the supply of ink.

Although necessarily more costly than the ordinary platen, in consequence of the greater amount of mechanical detail, the Napier, as before mentioned, possesses many obvious advantages. The inking is superior, inasmuch as the rollers pass twice over the forme— practically affording double-inking. Greater impression may be obtained at far less expenditure of power, and by the enlargement of the diameter of the grooved drum, the forme is at rest on the bed for a slightly longer time, allowing the platen to rise some little distance before the return is made. But multiplicity of movements has its disadvantages. Greater attention on the part of the workman is necessary in lubricating, &c.,

and unless the utmost care is exercised, the bill for repairs is liable to be excessive.

The makers of these machines have devoted a large amount of attention to the manufacture ; and inasmuch as they can boast of being veterans in the construction of printing machines, the greatest confidence may be placed in the engineering work, which is excellent. as both their old grippers and platens testify.

CHAPTER VIII.

PERFECTING MACHINES.

General Description of Dryden's and Middleton's—Various Modes
of receiving Sheet and Delivery—The "*Queen*," Dawson's New
Perfecting—Payne's Machine—Newsum's Perfecting Machine—
The Single-cylinder Perfecting Machine.

ALTHOUGH the Anglo-French and several others print
the sheet on both sides at one operation, in the trade
the Perfecting machine proper is generally understood
to refer to those made by Messrs. Middleton or Messrs.
Dryden and Foord. All perfecting machines of this class
are exactly similar in construction, but three different
motions are employed for carrying the sheet into the
tapes—the gripper, drop-bar, and web. The first-named
is generally preferred for accuracy of lay, the second
for speed, while the third, although still in use on some
machines, may now be considered obsolete, offering no
advantage over the gripper for precision, while it is
impossible to feed-in more than 1,000 sheets per hour,
added to which the work of the layer-on is of a much
more trying nature.

Briefly, the Perfecting machine consists of two large
cylinders, inner and outer, having intermediate drums
between for the turning or reversing of the sheets. The
side-frames are low, to enable the minder to attend to his
formes with ease. The main driving shaft, called the "lay-
shaft," is in the middle of the machine, on the off-side, and
drives both the cylinders and reversing motions direct.

The side-frames stand about two and a half feet high,

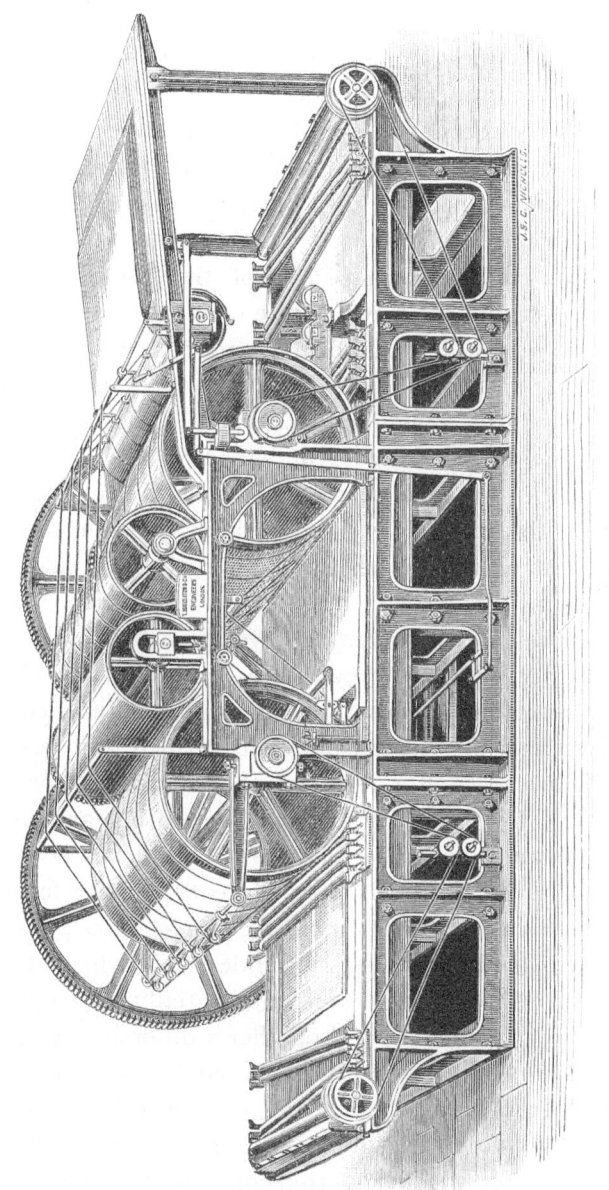

FIG. 41.—MIDDLETON'S FAST PERFECTING MACHINE

the length of course being dependent upon the size of the
sheet the machine is constructed to print. At equal distances from either end are placed the large impression
cylinders, the supports being substantial supplementary
frames or cross-girders resting upon the main side-frames.
The circumference of these cylinders is equal to the length
of the two coffins, which admits of the forme returning
the reverse way to the travel of the cylinder between every
impression. It will, then, be understood that only a portion of the surface is used for the impression—in fact,
about one-half, the remaining portion being much thinner,
to allow the forme to pass underneath without coming
into contact. To equalise the weight of the cylinders, a
counter-weight of iron is fixed inside, opposite the impression, calculated to be equal to the extra thickness
of metal on the thicker or impression portion. This,
of course, tends to balance the weight when travelling.
Between the impression cylinders, and slightly above,
are the reversing drums, one of which is utilised for
the adjustment of the register, and termed the register
drum. The sheet, after receiving the impression from the
inner forme by travelling over the first drum and under
the second, and on to the outer forme, becomes reversed,
and the unprinted side is then presented to the forme.

The coffins (two) and ink-tables (two) are provided
with well-planed bars on the under-side, which travel
upon a series of strong and substantially set pulleys.
These pulleys are made on altogether a different plan from
the runners employed on the Wharfedale, each being in
its own bearers and independent of the others. These
are fitted in two light girders running the entire length
of the machine, and supported on the end and cross
frames. Immediately under each of the cylinders are
placed two larger wheels or pulleys, called impression

pulleys (Fig. 42), affording extra strength at this point to insure a solid resistance.

The tables are moved to and fro by means of a pinion-wheel and rack—a most ingenious and successful combination. When this machine was first conceived by Donkin and Bacon, the greatest difficulty which presented itself was the reversal of the tables. When it is considered that the combined weight of the travelling

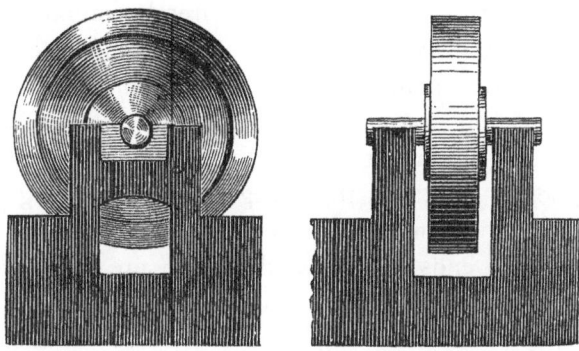

Fig. 42.—Impression Pulleys.

beds and the impetus are equal to several tons, to perform this successfully without excessive strain or wear upon any portion was certainly an achievement.

The main driving shaft extends to the centre of the machine between the two cylinders, and is fitted with a strong bevel-wheel working into a similar one at right angles. The base of the perpendicular shaft, called the upright spindle, is secured in a socket. At its top is a pinion-wheel, immediately above which is a substantial pulley. Upon the pinion and pulley is thrown the chief strain of the reversal of the tables, so they are both necessarily very solid in construction. An endless rack in the form of an elongated oval is fitted to the under-side of a large frame. The teeth lie in a perpendicular direction, and into them fits the pinion on the upright spindle. The pulley travels in a wide groove immediately above, and materially assists to steady the motion of the rack. At either extremity of the parallel series of cogs are the semicircular "ends," which are generally made of steel or malleable iron, and can be renewed with little trouble in case of necessity. These ends are more liable to be broken than the side racks, as the pinion, at a high rate of speed, travels directly round, or from one side to the other. An immense strain is necessarily exerted at this point, as when the pinion is at the extremity of the rack, the tables have been suddenly brought to a standstill, and immediately propelled in the opposite direction, the upright spindle having simply a rotary motion in its own collar.

The two ends of the rack frame are supported on the under-side of the tables by iron bars, between which they work freely, allowing a strictly steady motion from one side of the tables to the other when the pinion travels round the end. Although the rack is both heavy and large—varying from 3 ft. 6 in. to 5 ft. 6 in. long, to 2 ft. and 3 ft. wide (according to the size of the machine) —the whole moves exactly in a horizontal position from one side to the other at the same time. This is effected by the adoption of a parallel motion—consisting of a series of steel bars, so arranged that neither extremity can move horizontally alone. Without some contrivance of this kind, when the pinion travelled round the end, the inclination would be for that portion alone to be forced over, leaving the remaining part at an angle.

This parallel motion consists of two steel bars, connected in the centre by a cross piece. On opposite

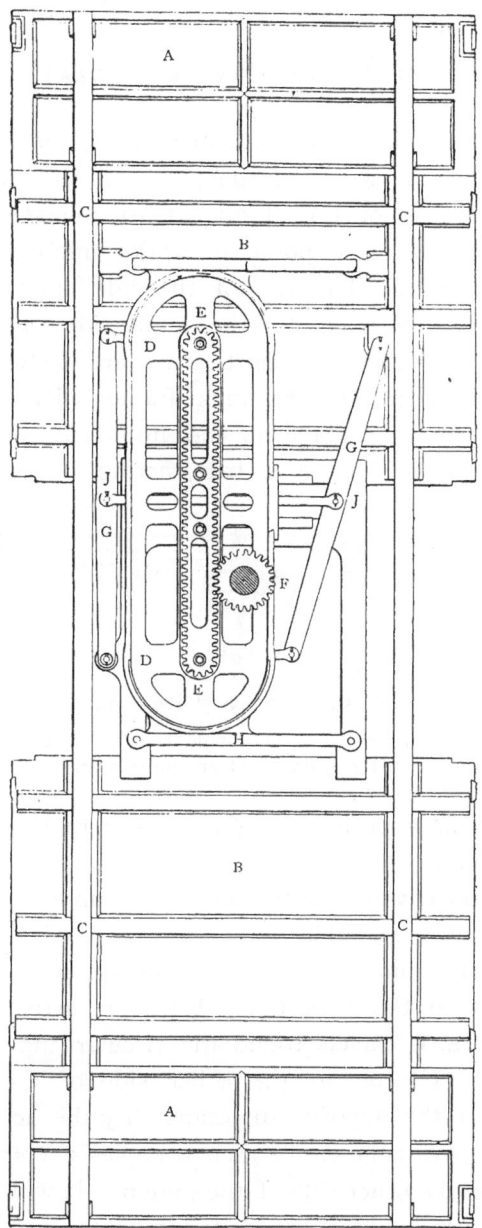

Fig. 43.—Under-side of Messrs. Middleton's Perfecting Machine, showing
Rack and Parallel Motion.

A A Ink-table; B B Coffins; C C C C Table bars; D D Rack; E E
Extremities of Teeth of Rack; F Pinion of upright Spindle; G G Parallel
motion; H H Fixed bars supporting rack; J J Bar connecting parallel bars.

ends of these bars are attached short pieces, each as
long as half the width of the rack. It should be stated
that the whole of this frame is so joined with loose
rivets, that at each point it works freely. We have
then four ends. The two of the side bars at opposite
extremities are fixed to the under-side of the tables,
while the ends of the short arms are fastened to the
centre of the top of the rack frame. By this means
the *entire rack* is held firm, and immediately the pinion
forces one end over, the bars, acting in concert, carry

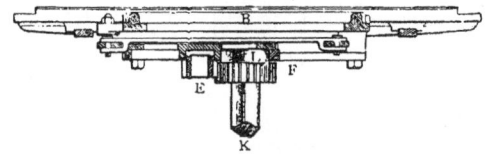

Fig. 44.—Cross Section of Rack and Table.

B Coffin; E Rack; F Pinion of upright spindle; K Upright spindle;
L Pulley on top of upright spindle.

the whole in a perfectly rigid manner to the other side
of the tables.

The above will perhaps be better understood on reference to Fig. 43. The pinion on the upright spindle F
is held fast in its bearings in the centre of the machine,
and works in the rack E E. When it arrives at either
extremity, E, it travels round the circular end and forces
the entire rack on to the other side of the machine,
moving it in the opposite direction. By the action of the
parallel bars G G the entire rack occupies a corresponding
position on the other side of the pinion. It will be noticed
that the parallel bars are bolted together in the middle
at points marked J J. The rack is supported on the
under-side of tables between bars at H H.

The cylinders are driven direct from the main shaft,

upon which is fixed a small pinion wheel working in gear with the inner-forme cylinder large cog. This latter works in gear with the outer-forme wheel. The intermediate or reversing drums are driven by a sup- plementary cog-wheel inside the outer-forme large wheel. These drums are sometimes made of wood, being sub- jected to no particular strain. One, the register drum, can be raised or depressed slightly on either side by means of screws, and assists the machine-minder in making register. The inking apparatus is similar to other machines—a duct being placed at either ex- tremity. The rotary motion is derived from a gut band working round a wheel on each of the impression cylinders. A series of graduated rings are so arranged as to enable the band to travel fast or slow—a corre- sponding wheel being fixed to end of duct roller—thus rendering any alteration in length unnecessary, as when it is put upon a small groove or cylinder it is allowed to run upon a larger one on the duct. That the bands may not interfere with the workman in attending to the forme, &c., they are taken down almost perpendicularly to the base of the machine round pulleys, and then up again to the ductor. These pulleys can be raised or depressed and secured by set-screws, thus enabling the minder to keep the band to the desired tension.

Compared with the Anglo-French, the rolling on this machine is not so perfect, as only three inkers entirely cover the forme. There are, however, four rollers, but the one nearest the cylinder "turns" on a full-sized forme. The distributing arrangements consist of four wavers, working in their forks at angles. The vibrators are worked by eccentrics from the cylinder shafts.

Endless set-off paper can be run on this machine, the only additional appliance necessary being a roller

fitted on brackets above the taking-off board, and near
to the outer-forme cylinder. The slips of paper are cut
into lengths, passed over and under the outer cylinder
and round the above-mentioned roller—being joined by
thin paste. The length of this endless piece of paper
being greater than the circumference of the cylinder,
insures a different portion presenting itself to the impression every time. This, however, rather concerns the
printing proper, which we shall deal with subsequently.

It will be noted that the taking-off board of this
machine lies under the reversing drums and between the
impression cylinders, so that the work cannot be watched
so conveniently as on other machines, where each sheet
is open to the light immediately it is thrown out; therefore
the minder usually is provided with a bank, or table, upon
which he frequently lays and examines the newly printed
sheet.

The tapes, of which there are two sets, called inner-
and outer-forme tapes, are of course endless. A diagram
to show the travel will be found on page 315. The two
sets travel together from the outer forme, where the sheet
is fed-in, round the inner cylinder, over and under the
drums, and over the outer cylinder. They separate at
the point immediately above the taking-off board—thus
releasing the sheet—the latter, possessing impetus, being
thrown upon the heap. One set travels over a tape-bar
down under and over the outer-forme cylinder above the
intermediate drums, when they travel again in company
with the inner-forme set. The second series continue
their course over tape-bar · and down under the inner
forme, up outside the cylinder under the laying-on board,
where they again meet the other series.

These tapes are so guided by pulleys that they
run strictly together, flush. The principal pulleys are

placed on the bars over the inner and outer formes, and consist of deeply flanged gun-metal wheels, working in a fork on a short, square spindle. At intervals along the bars are brackets, having a bore through which the bar passes on one side, while on the other is another hole into which the spindle of the tape-pulley fits. Set-screws are provided in both cases, so that the bracket can be moved to any portion of the bar, and the pulley adjusted up or down to enable the tapes either to be loosened or tightened. These are the only points where the tension can be regulated. On the other bars the tapes are either allowed to run without guidance, or indiarubber slings are used to assist them to travel straight. Fixed pulleys—which are small grooved wheels secured by set-screws in any position along the bars—are sometimes adopted, however.

Fig. 45.—Tape Pulley

As before stated, three different systems are adopted for receiving the sheet :—

The Gripper is now perhaps mostly used for this class of machine. An extra cylinder is provided immediately above the inner-forme cylinder, having a series of grippers fitted in an aperture, called the mouth of the cylinder, in the ordinary way. This cylinder revolves twice between each impression, taking the sheet every other time, and releasing it into the tapes on the underside. The gripper-bar extends to the outside of the drum, and has a " shape " attached, which is moved open by a cog or small cam on a cog-wheel. This latter is so calculated as to come into contact with the gripper-bar at every alternate revolution of the cylinder, closing the grippers upon the sheet laid to a front mark. Immediately below, on the side-frame, is a strong fixed pin, which forces the gripper-bar open, allowing the sheet to run into the tapes. The closing of the grippers on to the iron drum and the opening again is certainly a noisy operation.

The gripper motion is much to be preferred fo exactness of lay, as the whole lay-edge of the sheet is firmly held in position until securely passed into the tapes. They are slightly more expensive than the drop-bar, in consequence of the additional iron drum. The speed is generally between 1,400 and 1,700 per hour.

The Drop-bar.—The sheets are in this case run into the machine by means of discs or "bosses" on a revolving bar falling upon the front edge of the sheet. These discs are about one inch wide, and can be placed in any position and secured along the bar by the aid of small set-screws. At the extremity, on the near-side, a short arm with

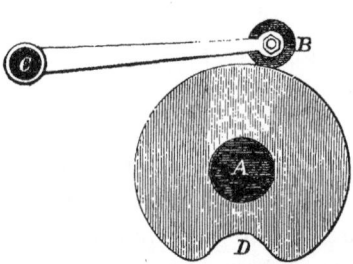

pulley B at end is fixed. This runs on a wheel with a "dip" D, thus allowing the discs on the bar to fall upon the sheet when the inner cylinder is ready to receive it. The bar is only momentarily in contact with the sheet, allowing the travelling tapes to complete the work.

Fig. 46.—The Drop-bar Motion.

A Shaft of disc; B Pulley; C Drop-bar.

It will be seen that the precision with which the paper is taken into the tapes is due entirely to the manner in which the bar does its work—always supposing, of course, that the layer-on strokes down the sheets exactly to the "marks." If the bar is slightly bent, or one of the discs a little worn, allowing the others to

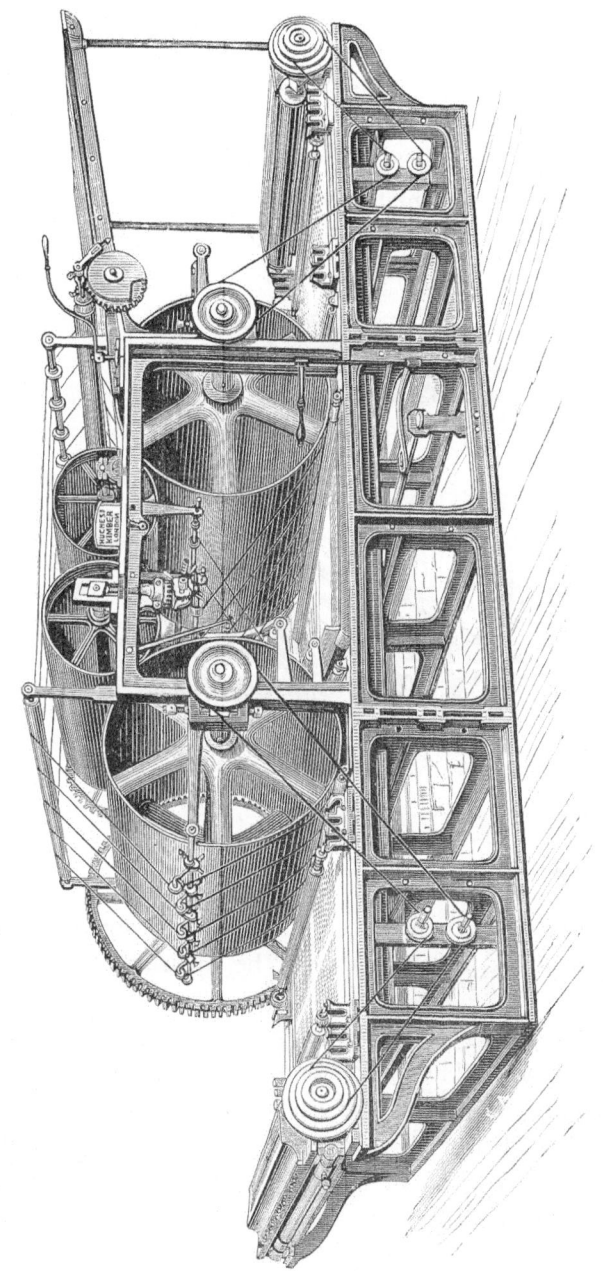

FIG. 47.—PERFECTING MACHINE, SHOWING DROP-BAR MOTION.

L

touch the sheet a second sooner, the sheet will certainly run into the machine slightly out of the square. Again, if the paper be wavy from any cause, the same defect will be apparent.

For speed, however, perhaps this mode of taking the sheet is to be preferred, as it is extremely simple

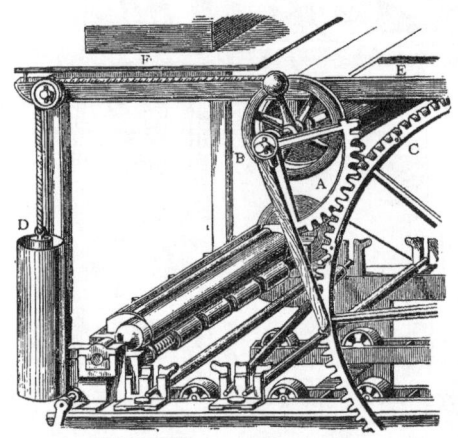

Fig. 48.—Web laying-on Motion.

A Segment ; B Web drum ; C Inner-forme cylinder wheel ; D Weight ; E Laying-on board ; F White paper.

and there is consequently very little to get out of order ; but for accuracy of lay it cannot be invariably depended upon.

The Web is so named because the sheet is laid upon a web of tapes. There are really two laying-on boards, perfectly flat, the one at the back, upon which the white paper is laid, being slightly raised. Immediately under the lay-edge is a small wooden drum, round which a series of broad tapes are fixed, continuing underneath the front laying-on board and over the surface, being in fact endless. These tapes are of course independent of the ordinary travelling series. On the end of the tape drum is a segment A, with teeth so made as to work in gear with a short series of supplementary teeth on the large inner-forme cylinder wheel C. Fixed to the under-side of this segment is a wooden arm, which extends beyond the outer flange of the cylinder wheel. On the latter is a strong pin, which, once every revolution of the wheel, comes into contact with the segment arm, forcing it down, and at the same time putting the segment into gear with the large wheel C. Immediately the last tooth of the segment is released the drum is pulled back to its original position by a weight D attached to a pulley on same shaft as the sector—the cord running parallel with the side-frame supporting laying-on board, and round a pulley down by one of the uprights, as will be seen in Fig. 48. Unlike the plan adopted in other machines, the sheet is laid to *back marks*, fixed on the back portion of the web tapes. When the sheet has been laid in position, the sector-arm being forced into gear gives the web-drum a forward motion. The tapes, being fixed to the drum, move in the same direction, and the sheet is thus fed into the machine. The return of the web, regulated by the weight, is rapid, to admit of the layer-on "flying" the next sheet into position.

In the case of the drop-bar and gripper motion, the sheets are usually "stroked" down to the front marks, which, while giving less work to the layer-on, enables him to feed it with great rapidity—1,400 to 1750 per hour. This speed is altogether out of the question with the web, as every sheet has to be taken up separately and "flied" over the web, and adjusted to the back and side marks.

L 2

THE "*QUEEN*" PERFECTING MACHINE.

This machine, named after the *Queen*, for which publication it was originally constructed, is made by Messrs. Davis (the manufacturer of the "*Graphic*") and Pardoe. Although at first sight it is suggestive of the Anglo-French, it is practically a perfecting Wharfedale machine. It differs, however, from the former in every respect as far as its construction is concerned. The two small impression cylinders work in gear with the table racks, and by an ingenious motion on either side, one cylinder wheel is alternately thrown out of gear as the tables reverse. When the outer forme is being printed, the cog-wheel on the cylinder is in gear with the rack, the corresponding wheel on the inner cylinder on the off-side running loose on the shaft. Immediately the tables are reversed, to allow the cylinders to still travel in the same direction, the outer cylinder runs freely back in the rack loose on its shaft, while the other side, being locked, drives both cylinders. Inasmuch as the cylinders depend for their motion on the table racks, it will be understood that they are necessarily stationary for a moment on the return.

In lieu of the ordinary rack and spindle, the tables are driven by connecting shafts, similar to the Wharfedale. The sheet is fed into the inner-forme cylinder in a similar manner to the Anglo-French, grippers of course being employed, which transfer it to the outer forme. No arrangements for set-off are made, but could be easily fixed if desired—on the same principle as that on the Anglo-French. The impression cylinders are alternately moved up and down by means of cams at the base of the cylinder frame. These cams, it may be mentioned, do not describe a complete revolution, but

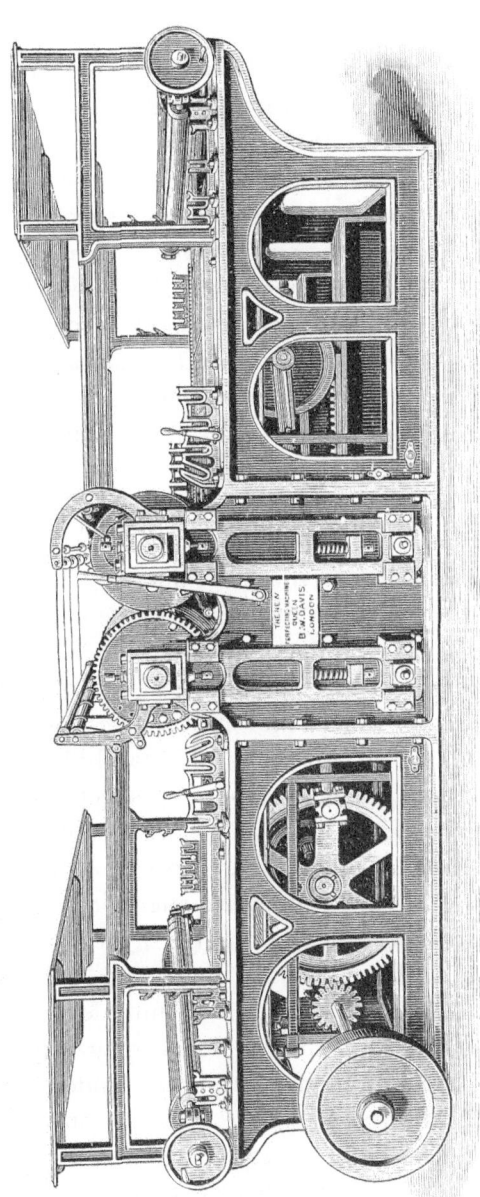

FIG. 49.—THE "QUEEN" PERFECTING MACHINE.

move only to the extent of a quarter of a circle—
backwards and forwards. There are no complicated
movements, the machine being singularly simple—which
is an advantage, as there is much less liability of
accidents, or of frequent repairs being necessary.

The machine is excellently made, special attention
being given, as in the case of the "*Graphic*," to the
supports of the table—three stout bars running under
the entire length—with extra support immediately beneath the cylinders. The maker maintains that no
cylinder packing is necessary, as the whole supports are
so substantial that it is impossible for a "dip" to occur.

DAWSON'S IMPROVED PERFECTING MACHINE.

In its main features this machine is identical with
the ordinary perfecting, the improvements introduced,
however, giving it a distinct advantage. The laying-on
board is lowered to the level of the top of the outer-
forme cylinder, to which the sheet passes over a small
drum. The taking-off apparatus is certainly both effective
and novel. By a series of tapes the sheet is conducted
from the outer-forme cylinder across the space the
taking-off ordinarily occupies, and passes over the first
register drum, and back round a smaller special roller over
the second register drum on to the flyer, which carries
it over on to a sloping board immediately in front of
the laying-on board, depositing it in full view of the boy.

The space between the cylinders may remain quite
open, so that the formes are practically under the eye of
the minder. Two inkers are placed on the frame
between, and nearly under each cylinder, thus insuring
good inking at the entry, which portion of the forme
in these machines generally suffers. It is advisable to

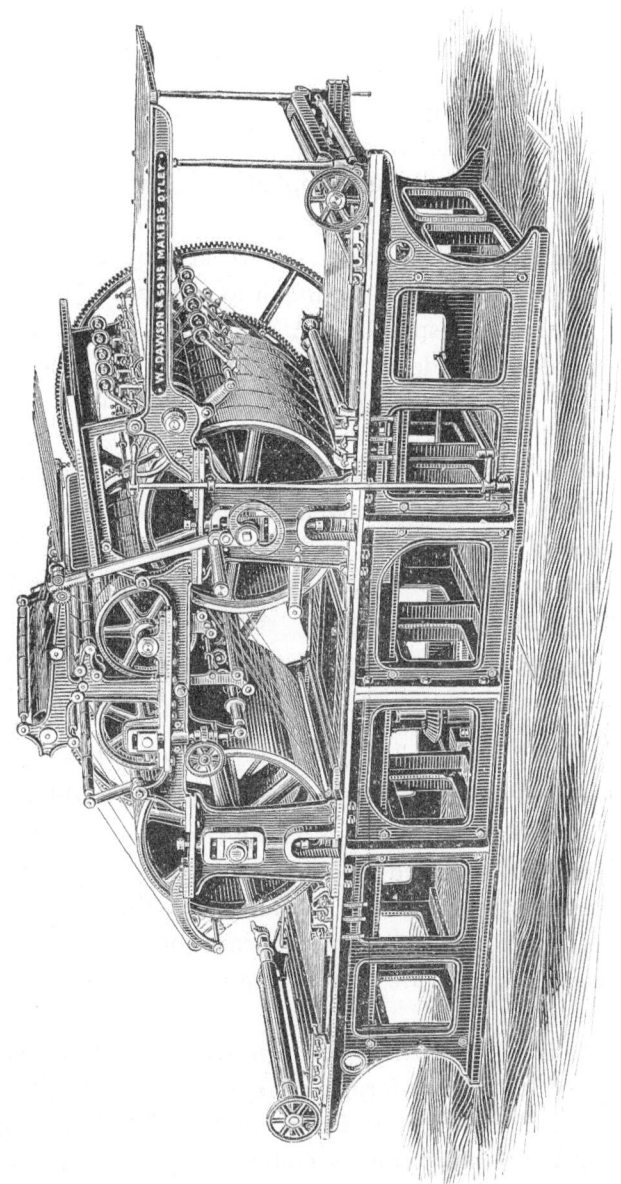

FIG. 50.—DAWSON'S IMPROVED PERFECTING MACHINE.

have a narrow board, however, laid across from one side to the other, as a serious accident might arise from falling in while the machine is in motion. The table of this machine is made in one piece.

The upright spindle is adopted to drive the tables, and the cylinders work in gear, as is usual with this class of machine. The flyers are worked by an upright rod with teeth at the end fitting into a geared wheel or the end of the flyer spindle. A large cam on the inner-forme cylinder, outside the frame, imparts the necessary movement. The register drum is regulated by a screw underneath instead of above.

It will be seen that the additions are really valuable, and wherever the Dawson machine has been worked, it is pronounced a success. It will do good work at 1,500 copies per hour, and of course requires but one boy—the layer-on.

PAYNE'S LARGE-CYLINDER PERFECTING MACHINE.

Messrs. Payne's perfecting machine varies considerably from the ones previously described. The large impression cylinders are close together, and working in gear, the cog-wheels being comparatively small and compact. This is a decided improvement, as it enables the machine-minder to work on his forme on the off-side, a matter of great difficulty in the case of large wheels. The register drums are placed immediately above the impression cylinders.

The laying-on board is fixed nearly on a level with the base of the inner-forme cylinder—no stage being necessary, and the sheet is fed-in to the grippers, which hold it until the impression is completed, an arrangement which commends itself as giving greater facility for accurate lay. The sheet is released from the inner forme into a series of tapes on the register drum, and taken by the grippers of the outer-forme cylinder, and

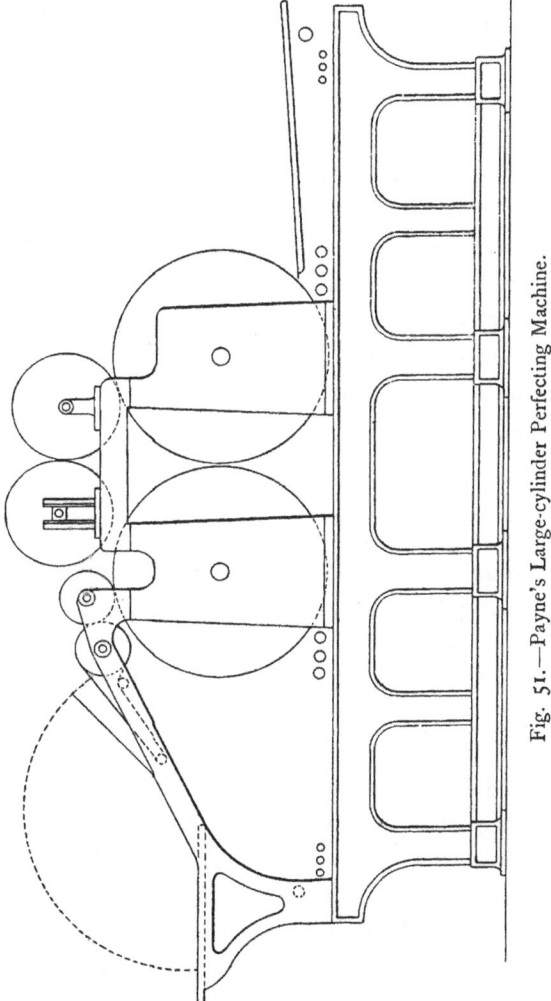

Fig. 51.—Payne's Large-cylinder Perfecting Machine.

finally deposited on to a set of flyers at the extremity of the machine. The printed sheet, therefore, unlike that on the ordinary perfecting, is exposed.

The tables are driven by the ordinary upright spindle and racks, and the whole of the gearing is machine-cut, imparting a solid and steady motion. The inking arrangements are almost identical with the same firm's Wharfedale.

The advantages possessed by this simple and in-geniously constructed machine are several. It requires less space both in length and height. The minder's labour in making ready is materially lessened by the cylinder being exposed ; the lay is unquestionable, and the successful application of the taking-off motion is certainly an accomplishment. The weak feature of the ordinary perfecting machine is undoubtedly the position of the taking-off board, where the ' work cannot be scanned as it is thrown upon the heap.

NEWSUM'S PATENT IMPROVED PERFECTING MACHINE.

This machine resembles the "*Queen*" in many respects —in its being an adaptation of the main features of the Wharfedale to the Perfecting machine. The impression cylinders are driven separately by the side racks, and the tables are propelled by a crank from the driving shaft attached to rack wheels, thus dispensing with the up-right spindle and rack. This is a greater advantage than it may at first appear, for the wide space usually occupied by the rack in the centre of the tables can have no support, as the rack must have free and unin-terrupted play. Then the additional runners which the application of the rack wheels enables the maker to add to this class of machine, give it additional value to printers requiring to work heavy and solid formes.

The most novel feature of Newsum's machine, however, is that it may be used either to print two distinct formes on one side, or to perfect. Two laying-on and taking-off boards are provided, and by throwing a centre cam out of gear, the cylinders become independent of one another, and deliver the sheets on to their respective taking-off boards.

This may be understood upon reference to Fig. 52. 1 and 10 are the cylinders working independently of one another, and in gear with table racks; 11 is the inner forme, and 13 the outer; 3 and 4 the laying-on boards, and 5 and 6 the taking-off. When worked as a per-

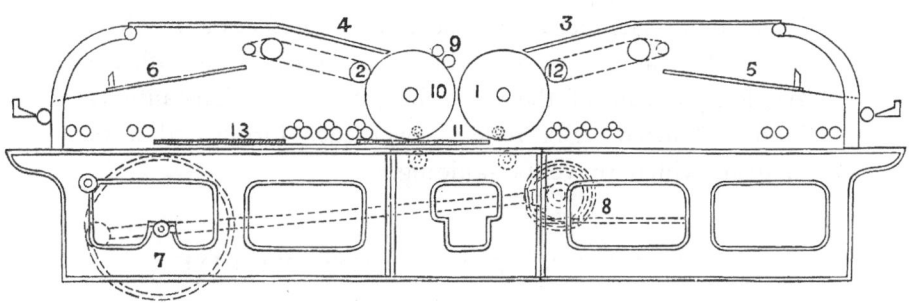

Fig. 52.—Newsum's Perfecting Machine.

fecting machine, the sheet is fed on to No. 1 cylinder, and after printing is released to the other at the point of contact, similar to the Anglo-French. It then passes over the tape frame 2, and is deposited on taking-off board 6. By simply altering a special cam, the grippers hold the sheet until the set of tapes 2, 12, under the laying-on boards, are reached, when it is released and deposited on its heap. The pair of set-off rollers 9 are also thrown out of gear when the machine is printing only one side; while when perfecting, the inner taking-off tape 12 may as easily be disconnected.

The advantage of printing two distinct jobs at the same time is problematical. No time is saved in the making ready, as when one forme is prepared it must remain until the second is finished, unless the minder runs the machine at half its capacity while he is making ready on the other end. In this case his attention is divided, and unless he is printing common work not requiring close attention, the delay in the making ready of the other forme must necessarily be great. The register of the sheet in perfecting must be dependent upon the lay, as there are no points provided. The idea is certainly most ingenious, but we anticipate that it will be found more satisfactory to use it as a perfecting machine rather than as two single-cylinder machines. As with the Anglo-American of the same firm, it is most substantially built and fitted. Six rollers cover each forme and work in gear, which prevents skidding.

SINGLE-CYLINDER PERFECTING MACHINE.

This machine is one of the most ingenious inventions of recent years. Originality is always to be admired, but in this case the innovation is so daring, and altogether so successful, that Messrs. Buxton, Braithwaite, and Smith are to be commended for introducing such a sound novelty. In general appearance it resembles a double-ended Wharfedale, or a two-colour machine with an extra-large cylinder.

Briefly, the cylinder is provided with a double set of grippers, one set being placed in the ordinary position at the surface of the laying-on board, and the other half-way round, *i.e.*, slightly over the extreme top. The surfaces between these grippers are each equal to the size of the forme that the coffins accommodate, so that both inner and outer formes have their special impression surface, answering in every respect the purpose of the

FIG. 53.—SINGLE-CYLINDER PERFECTING MACHINE.

second cylinder of the perfecting machine. The sheet is laid in the ordinary manner to the first set of grippers, and the impression is given by the inner forme. The paper then travels half-way round the cylinder, when it is released into a set of tapes fixed into the supplementary drum on the top. Thus the sheet is reversed, *i.e.*, the printed side turned downwards, and run on to the second laying-on board. When this movement takes place the front of the laying-on board is necessarily a few inches above the top surface of the impression cylinder. Immediately, however, the sheet is wholly deposited on the board, the front edge of the latter drops so as to bring the paper exactly in position for the grippers on the impression cylinder to grasp it. A finger on the side automatically moves the edge to a fixed side lay, so that perfect register is insured. After the second side is printed the sheet is conducted on to a set of flyers, and deposited on the taking-off board.

The inking is excellent, as many as six inkers being available. The riders are geared, working in an independent wheel outside the side-frame. Two fly-wheels are provided—one on each side of the driving shaft.

As far as the general arrangement of the working parts is concerned, they are almost identical with the Wharfedale. A centre girder is fixed under the cylinder to give the necessary solidity to the impression. If desired, the formes may be double inked, an advantage in heavy cut-work. The makers claim that first-class printing may be done at a speed of 1,200 copies per hour.

Unlike the ordinary perfecting, these machines require no pit, so that they may be erected on floors.

Machine-minders will appreciate the convenience in overlaying. The surfaces of the cylinder are thoroughly exposed, and the pasting-up is therefore very much easier.

CHAPTER IX.

THE ANGLO-FRENCH MACHINE.

General Description of Messrs. Marinoni's and Messrs. Hopkinson
and Cope's Anglo-French Machines.

IN general construction this machine is in many respects
similar to the Napier Gripper—a machine very much in
favour some years ago, but now nearly obsolete. It
differs from the ordinary perfecting, inasmuch as it
possesses but two cylinders, the sheet passing directly
from the inner on to the outer forme, the intermediate
drum for reversing the sheet being dispensed with. A
separate apparatus is also provided for set-off sheets,
the laying-on board being fixed above the outer-forme
cylinder, and single sheets are fed in to meet the partly
printed paper upon its entry on to the outer cylinder.

The cylinders work in gear on the middle of the side-
frames, but are really independent of the same, being
supported by two steel upright movable frames on either
side. The tables travel to and fro beneath, the cylinders
alternately rising and falling, as the inner or outer forme
is being printed. This is necessary to allow either forme
to clear the blanket when returning after the impression,
as, unlike the ordinary perfecting, the entire surface is of
nearly equal thickness, and of nearly the same circumference as the length of one coffin. The tables are propelled by means of a rack having horizontal teeth, being
driven by a pinion wheel from the driving shaft.

The Cylinders, as before mentioned, are supported

FIG. 54.—THE NAPIER GRIPPER : THE ORIGINAL OF THE ANGLO-FRENCH MACHINE.

(It will be noted that the inkers are placed immediately under the ductor, the distribution being accomplished by an intermediate roller.)

M

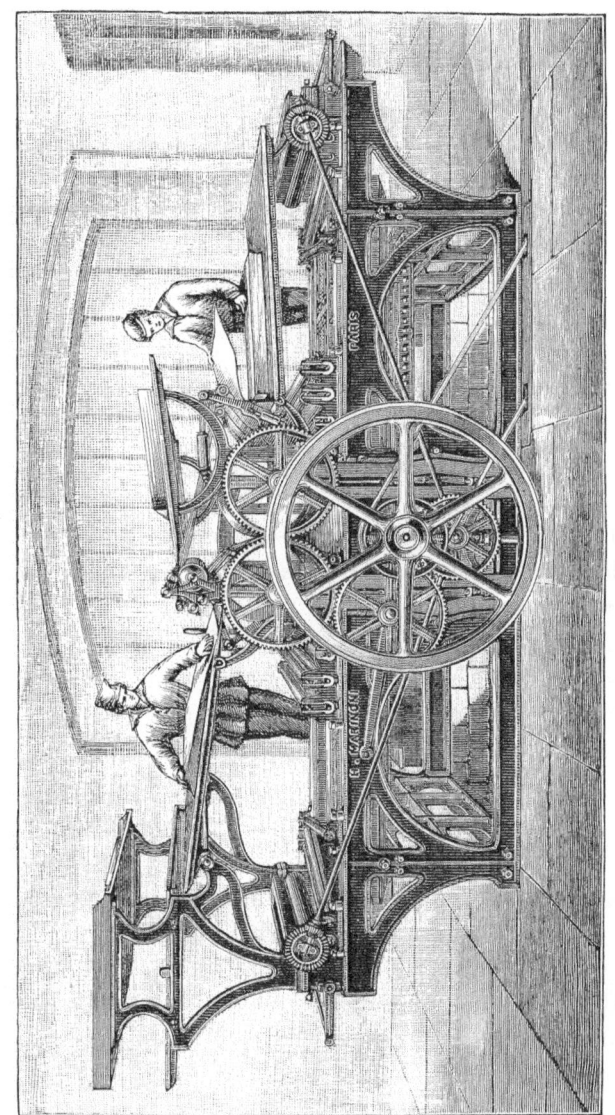

Fig. 55.—THE ANGLO FRENCH MACHINE.

upon a movable frame on either side.   The entire surface
is used for the impression, the blanket being sometimes
stretched entirely round, in a somewhat similar manner to
the Wharfedale.   Large cog-wheels are fixed on the cylinder
shafts on off-side, and work in gear.   The motion is imparted by means of an intermediate cog-wheel, working
in gear with a smaller wheel on the main driving shaft.
This will be readily understood by reference to Fig. 57.
E is the main driving shaft, and the wheel A, by working
in gear with the intermediate cog B fixed on side-frame,
drives the inner-forme cylinder C, which of course moves
the outer-forme cylinder D.   The travel of the various
wheels is indicated by the arrows.

Fig. 58 gives a sectional view, and shows the working
of the above series of wheels in conjunction with the
shaft attached to the universal joint, which moves the
tables to and fro.   A is the small fly-wheel, which is
of use in making ready, "backing up," &c., as well
as for the impetus it affords the machine in the running; B the bracket holding the main shaft; C live
and dead riggers; D is a small pinion driving a larger
wheel E; the shaft of the latter, extending under the
machine, works, by means of a powerful cam at J, the
rocking frame (hereafter described), which moves the
cylinder frames alternately up and down; K is the bearing supporting the end extremity of the cam shaft; F
the cog on the main shaft, which extends through the
frame to the universal joint L, one of the most ingenious
motions used in printing machines.   It will be seen that
from a bar in the centre of L extends a shaft terminating at the pinion-wheel M.   The main shaft of course
imparts a rotary motion to the shaft P.   The teeth of
the pinion M work in a series of teeth on a single
rack fixed on the under-side of the table, and itself being

firmly held in the collar
Q, propels the rack to its
extremity, when it travels
up into the position in-
dicated by the dotted
lines, and thus reverses
the tables, the pinion
being on the *top* of the
rack instead of *underneath*.
This perhaps will be
better understood on reference to Fig. 61 : A is an extension of the main shaft
immediately inside the
side-frame. The universal
joint is composed of three
separate pieces — the
bracket on end of A, the
collar C, and the inde-
pendent shaft B. A is
held firm in its bearing.
The studs 1 and 2 are
fitted into the slots 1 and
2, the collar C being
secured by the screws.
The projecting studs 3
and 4 are placed through
the corresponding slots
on bracket of B, and also
screwed up. We have
thus a firm shaft at A, but
the collar works loosely
at points 1, 2, 3, and 4,
allowing the shaft B free

M 2

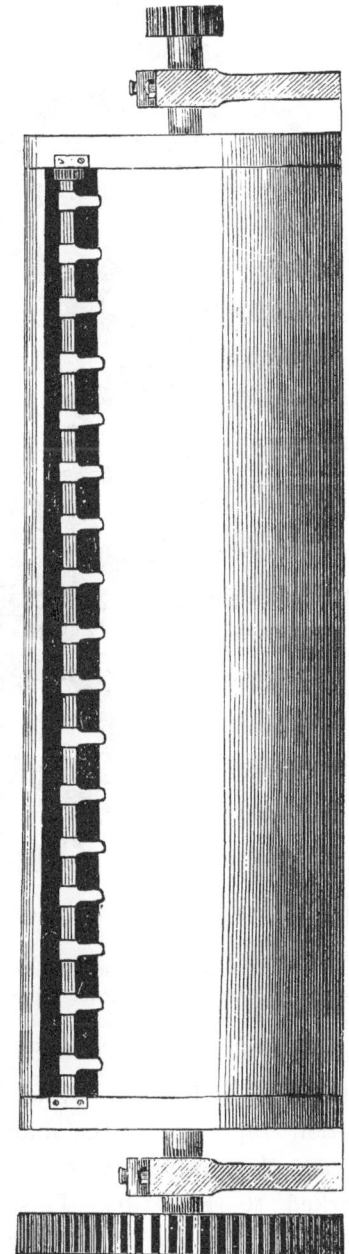

Fig. 56.—Cylinder of Anglo-French Machine.

Fig. 57.—Train of Wheels showing the Driving of the Impression Cylinders, &c.

A Cog on main shaft; B Intermediate wheel; C Inner-forme cylinder wheel; D Outer-forme cylinder wheel; E Driving shaft; F Wheel with extended shaft working rocking frame.

action up and down, while it still possesses the rotary motion imparted by the main shaft A.

At the extremity of the shaft P works a pinion marked M (Fig. 58). This pinion travels in gear with the teeth in the rack (Fig. 62). It will be noticed

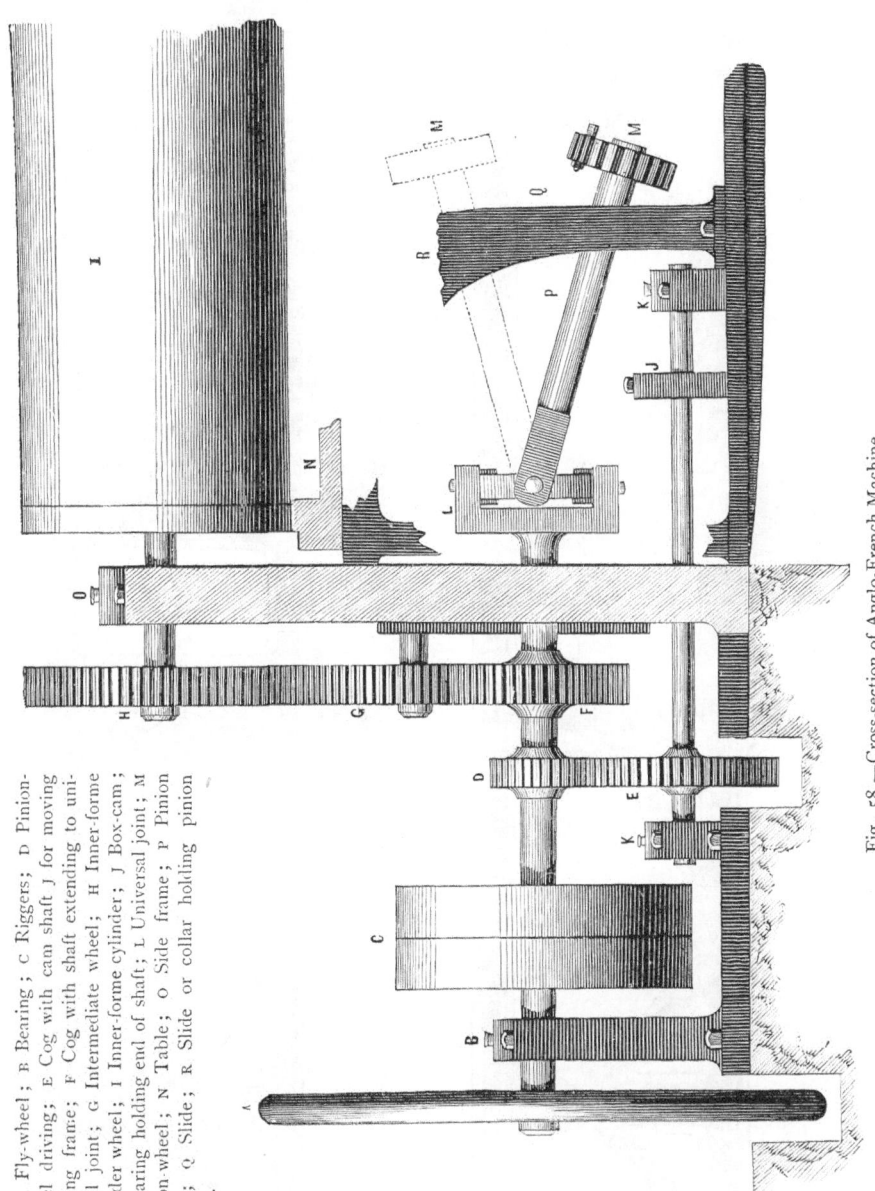

A Fly-wheel; B Bearing; C Riggers; D Pinion-wheel driving; E Cog with cam shaft J for moving rocking frame; F Cog with shaft extending to universal joint; G Intermediate wheel; H Inner-forme cylinder wheel; I Inner-forme cylinder; J Box-cam; K Bearing holding end of shaft; L Universal joint; M Pinion-wheel; N Table; O Side frame; P Pinion shaft; Q Slide; R Slide or collar holding pinion shaft.

Fig. 58.—Cross-section of Anglo-French Machine.

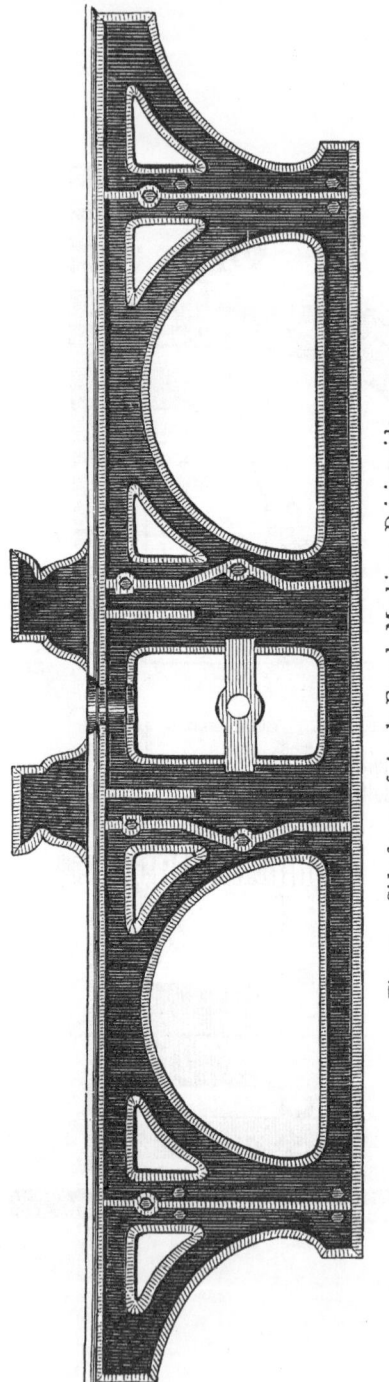

Fig. 59.—Side-frame of Anglo-French Machine : Driving-side.

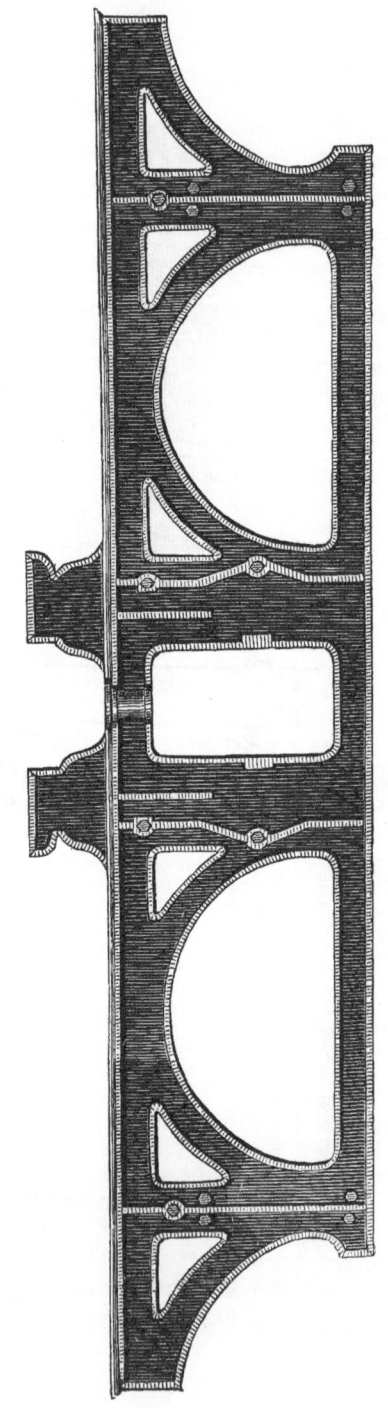

Fig. 60.—Side-frame of Anglo-French Machine : Near-side.

that the whole of the rack is fixed by strong brackets
to the under-side of the tables M M at the points I J,
there being, of course, sufficient space provided between
the cogs and the table to allow of free run. The

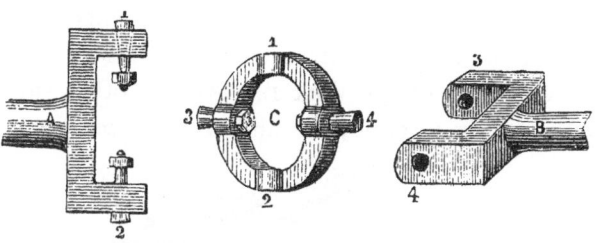

Fig. 61.—Shapes constituting the Universal Joint.

extent of the table is reached in three revolutions,
marked 1, 2, 3. When at the end F the pinion-wheel
moves rapidly round the steel shape H, the axis being

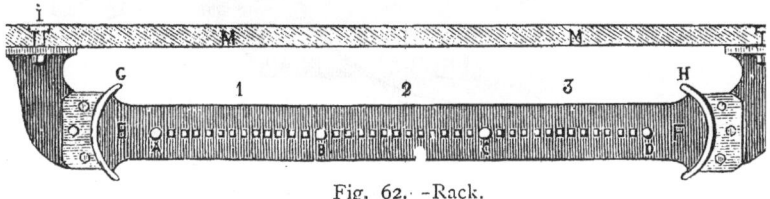

Fig. 62. -Rack.

A B C D Extra large teeth; E F Extremities of rack, up and down to
which points the pinion-wheel travels; G H Steel quadrants; I J Bolts
securing rack to under-side of table; M M Table.

at F, and thus forces the rack to travel in the opposite
direction, having changed its position from the top to
the bottom of the rack, but itself rotating in the same
direction.

It is obvious that when the pinion reaches the extremity of the rack there is a very severe strain, as the
tables are suddenly brought to a standstill and reversed.
There is not only the dead weight to be encountered,

but the impetus has to be destroyed in one direction and reimparted in another. This is really done by literally jamming the pinion-wheel between the end tooth D (Fig. 63) and the steel flange of the quadrant. It will be noticed that one cog of the pinion-wheel is slightly larger and of different shape from the others. This falls on the end stud of the rack D. At the other side of the pinion-wheel is a pulley K, and as the pinion is about

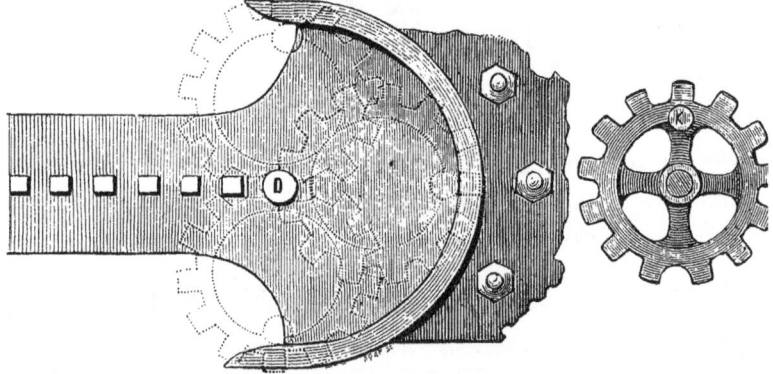

Fig. 63.—Portion of Rack, showing the Travel of the Pinion-wheel from one Side to the other.

B Pinion-wheel; K Pulley; D End stud of rack.

to descend it travels round the flange of the quadrant, the rack being held stationary between the end tooth D and the pulley when the pinion is in a line with the rack, indicated by the centre dotted wheel. As the pinion drops underneath, the rack is reversed, and travels until the other end is reached, when it runs *up* the quadrant in a similar manner. It not unfrequently happens that the end stud is snapped in consequence of the sudden jar. It is, however, generally screwed into the rack from the other side, and thus easily renewed. Three special cogs are fixed at intervals along the rack (Fig. 62). These are exactly similar to the end stud, and the same teeth of the pinion that turn on the extremities fall in these positions.

The return of the tables is assisted by two large upright buffer springs at either end, with which the under-side of the table comes into contact immediately before the pinion reaches the end of the rack, and the impetus is thus slightly reduced, while the return is assisted by the rebound of the springs to their position.

The rise and fall of the pinion shaft is regulated by the collar R (Fig. 58). Just sufficient space is allowed on either side to allow free working, while at the bottom the shaft is supported as in a bearing. When the pinion is travelling at the top of the rack, it of course rests on a strip of smooth iron half an inch wide, and running the whole length of the rack, over and close to the roots of the teeth.

The rocking frame is driven by the cam at J (Fig. 58) —a shaft below the main shaft—K being the bearings at each extremity. This frame consists of two strong parallel bars extending under the machine from one side to the other, the extreme ends in each case forming the lower portion of the

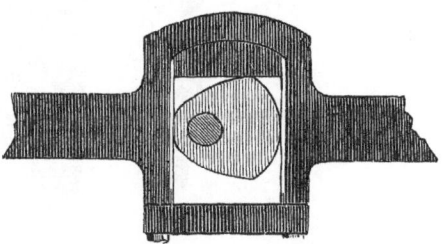

Fig. 64.—Cam working Rocking Frame.

knuckle-joint. In Fig. 65 H H H is the cylinder frame supporting the cylinder; D is the end of the rocking frame, which is the actual motive power; A is a steel bracket on the side-frame, and firmly secured, the short steel arm C fitting in the slot of A at the top and into the rocking frame end D. It will be understood that when D is moved into a perpendicular position, the

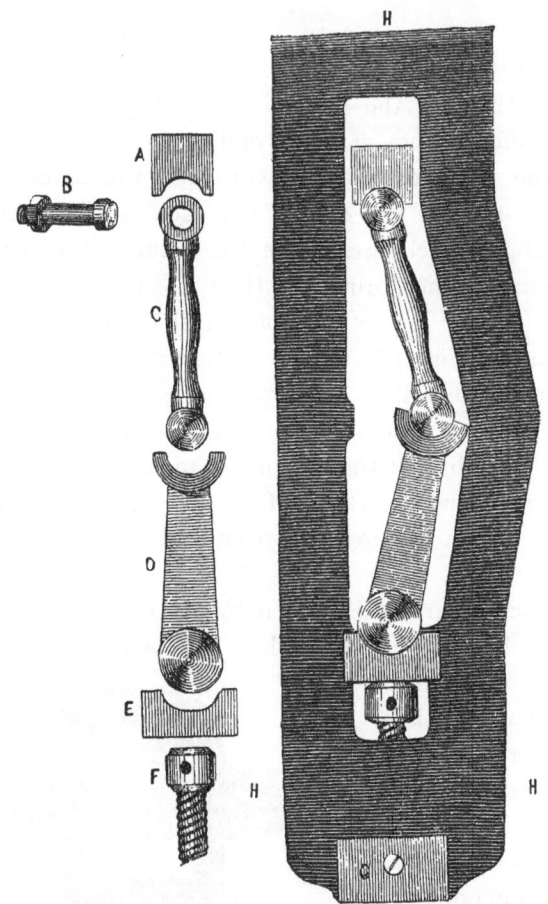

Fig. 65.—Levers at Extent of Rocking Frame, for raising and depressing
Cylinder.

A Fixed point on side-frame; B Bolt; C Loose steel shape; D Extension of
rocking frame; E Bearing; F Impression screw; H H H Cylinder frame.

bracket A, being fixed, will force the frame H H H (carrying
the cylinder) downwards. When in the position shown,
the frame (and consequently the cylinder) is at its highest,
allowing the forme to pass underneath without touching
the blankets which are going round with the cylinder in

the reverse direction. By means of the screw F, which can be raised or depressed, the impression is regulated. It will be noticed that the whole of this motion is made up of nicely adjusted parts.

A   B

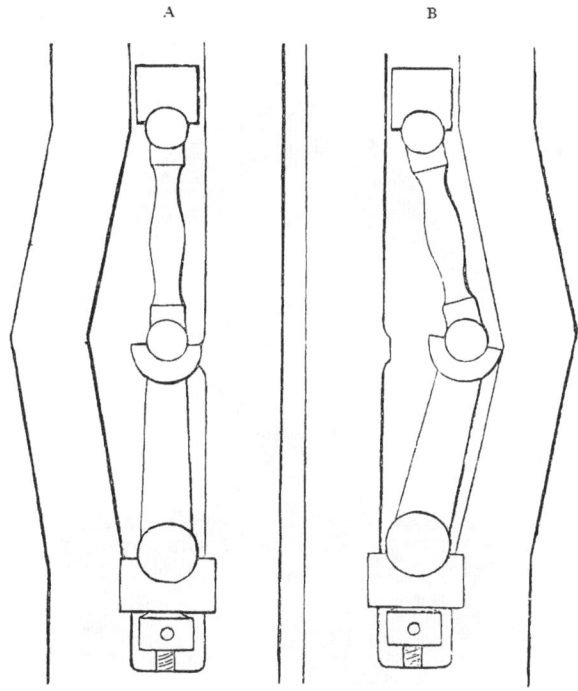

Fig. 66.—Cylinder Frames, showing relative Position of Rise and Fall Motion when Impression is being taken.

Each cylinder has a supporting frame on either side, which of course moves simultaneously, the portion of the rocking frame corresponding with D extending to the other side of the machine. The cylinders are never actually on a dead level, but rise and fall sharply, alternately. Fig. 66 shows their relative position. It will be noticed that the joints at A are perfectly rigid, while at B they have followed the shape of the cylinder frame. This will be the actual position when the cylinder supported by A is taking the impression, while the other cylinder on B is slightly raised. Immediately the table returns the joints of A fall into the angle, and B becomes at the same time straight, thus enabling the impression to be given by the second forme.

Fig. 67.—Bracket or Side-frames for steadying Rise and Fall of Cylinders.

A Supporting bracket; B Knuckle; C Box spring.

The cylinders are of course of some considerable weight, and if the rise and fall depended solely upon the action of the cam of the rocking frame, the motion would be apt to be unsteady. To obviate this a strong spiral spring is placed in a box on the top of the side-frame, between the cylinders supporting a steel shape with slots on either side (Fig. 67, A). These latter bearings fit into a shape on the inside of each of the cylinder frames, and when the cylinders are forced up or down impart the desired steadiness of motion. It will be noticed that it works upon a knuckle at B, and so can move easily up or down at either extremity. A section of the spiral spring is also shown in the same diagram. The amount of support can be regulated by the set-screw.

The grippers of both the cylinders are governed by

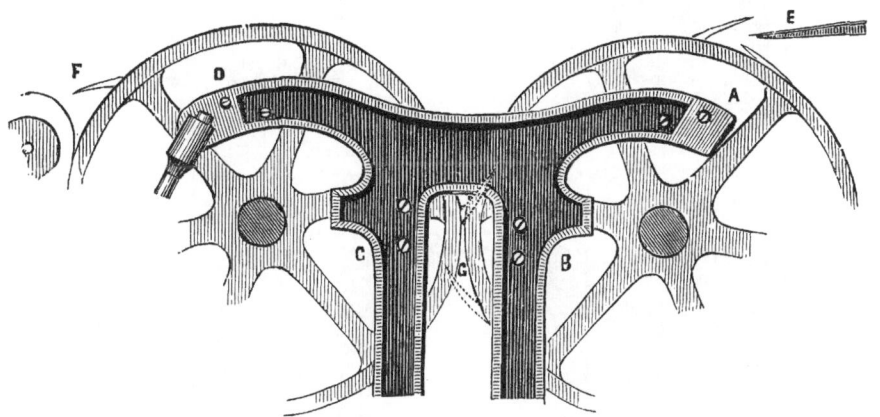

Fig. 68.—Folding Frame for opening and shutting Grippers.

a series of shapes upon the folding frame on the near-side of the machine (Fig. 68). Two bars extend close to the side of either cylinder, and the shapes, fixed on the inside, are so arranged as to come into contact with the tumblers attached to the gripper-bars on the out-side of the cylinder. A small cog-wheel on the end of the outer-forme cylinder shaft works into another cog. On the inside of the latter is a flange, with a dip. On this travels a small pulley fixed to the side of the folding frame, and by this means the frame upon which the gripper shapes are fixed is forced in or out. When the folding frame is close in, the pulleys at the

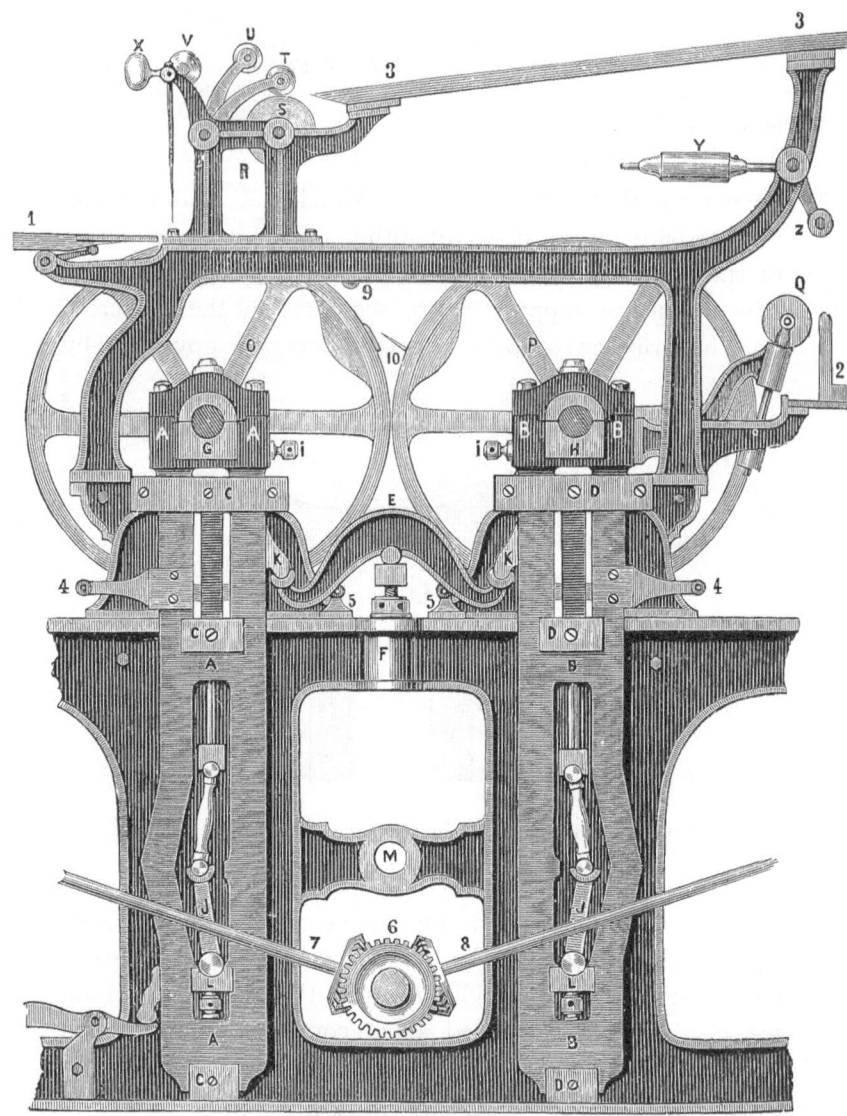

Fig. 69.—Showing Motions, &c., on the Driving-side of Machine.

A Outer-forme cylinder frame; B Inner-forme cylinder frame; C D Steel guides; E Bracket for steadying cylinders; F Box spring to steadying cylinders; G H Cylinder brasses; J J Chill of rocking motion; L L Steel shape supporting bolt of ditto; K K Shapes on cylinder frame working loosely in spring bracket; M Bearing for universal joint shaft; O Inner-forme cylinder; P Outer-forme cylinder; Q Tape pulley; S Set-off roller; T Drop-bar; U V Tape pulleys; X Balance weight; Y Balance weight; Z Tape pulley; 1 Laying-on board; 2 Taking-off board; 3 Set-off laying-on board; 4 4, 5 5 Tape bars; 6 Cog driving ductor rods; 7 8 Ductor rods; 10 Grippers.

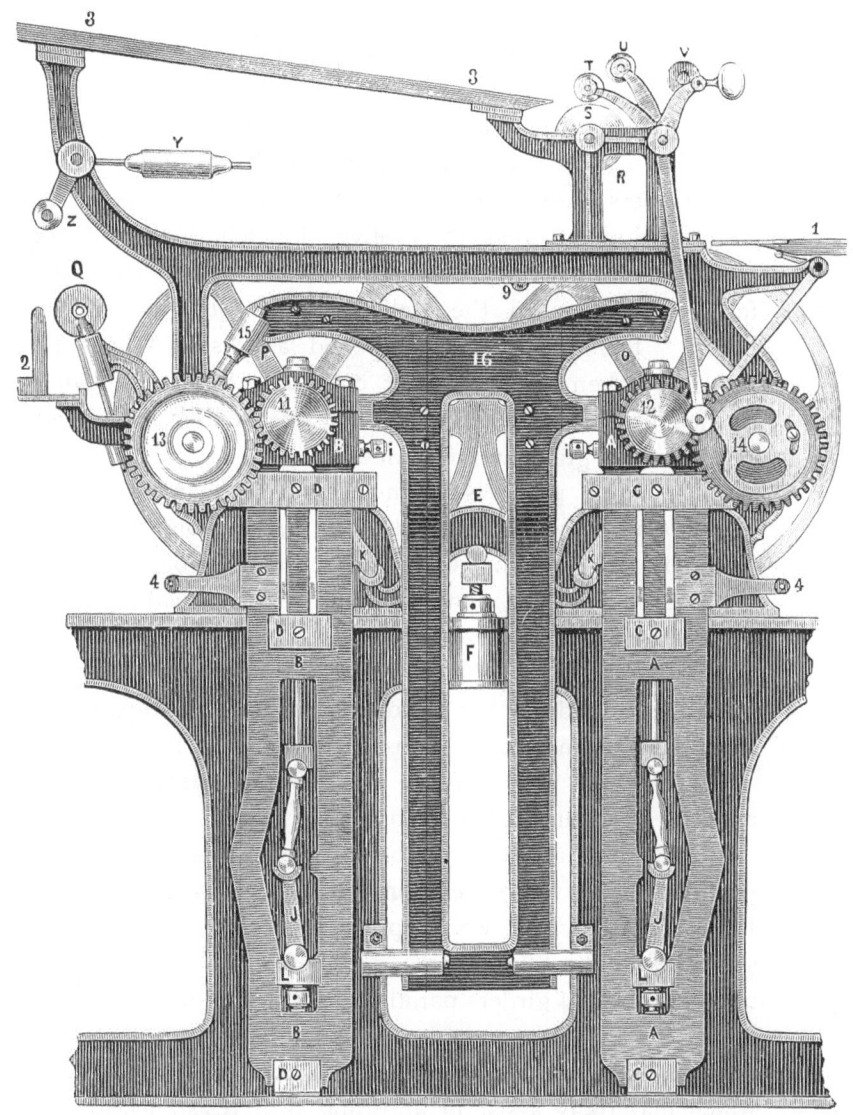

Fig. 70.—Showing Motions on the Near-side of Machine.

A Outer-forme cylinder frame; B Inner-forme cylinder frame; C D Steel guides; E Bracket for steadying of cylinders; F Box spring to steadying of cylinders; K K Shapes on cylinder frames working loosely in spring bracket; O Outer-forme cylinder; P Inner-forme cylinder; Q Tape pulley; S Set-off roller; T Drop-bar; U V Tape pulleys; Y Balance weight; Z Tape pulley; 1 Laying-on board; 2 Taking-off board; 3 Set-off laying-on board; 4 4 Tape bars; 11 Cog on cylinder shaft in gear with wheel for regulating folding frame; 12 Cog on cylinder shaft in gear with wheel for working set-off drop-bar; 13 Wheel with inside shape in connection with folding-frame pulley; 14 Shape for working set-off drop-bar; 15 Bracket with pulley on extremity of folding frame; 16 Folding frame.

end of the gripper-bars glide over the shapes, and open momentarily, their own springs inside the cylinder closing the grippers immediately the shape is passed.   The inner-forme grippers open to receive the white paper, and almost simultaneously the outer forme releases the per-fected sheet on to the delivery tapes just below the board.   The two sets are opened and shut when the sheet is taken by the outer cylinder ; both sets being opened at the same time, one to release the sheet, and the other to grasp it for perfecting.

The set-off paper is taken in by a revolving drop-bar. Corresponding with the small cog driven from the outer cylinder, a similar cog works from the shaft of the inner On the outside is a flange or cam, upon which works a pulley at the end of an arm extending from the bar immediately above the cylinders.   When the pulley reaches the "dip" the revolving discs drop upon the sheet laid to marks, which is carried by an independent series of tapes to meet the half-printed sheet as it is entering the outer-forme grippers.

The coffins are unlike those of the ordinary perfecting machine, being separated simply by a narrow bar (see Fig. 72).   They are well supported on either side by a series of runners fitted into parallel bars.   These travel upon substantial girders parallel with the side-frames.   Im-mediately under the cylinders large impression pulleys are provided, consisting of solid wheels about two inches wide, which materially assist in affording firm resistance to the cylinders when impression is being taken.

In Fig. 73 we give a sectional view of the table im-mediately under the cylinder.   It will be noticed that the beds are well supported by the runners on either side C C and the impression pulleys D D.   In the centre E is the rack, fixed to the tables by counter-sunk screws B.

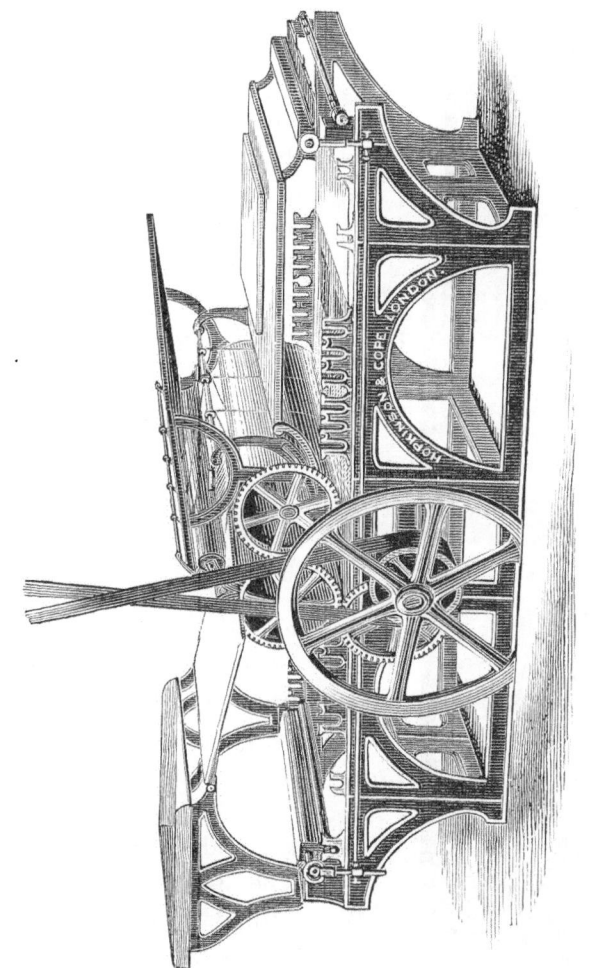

Fig. 71.—HOPKINSON AND COPE'S ANGLO-FRENCH MACHINE.

N

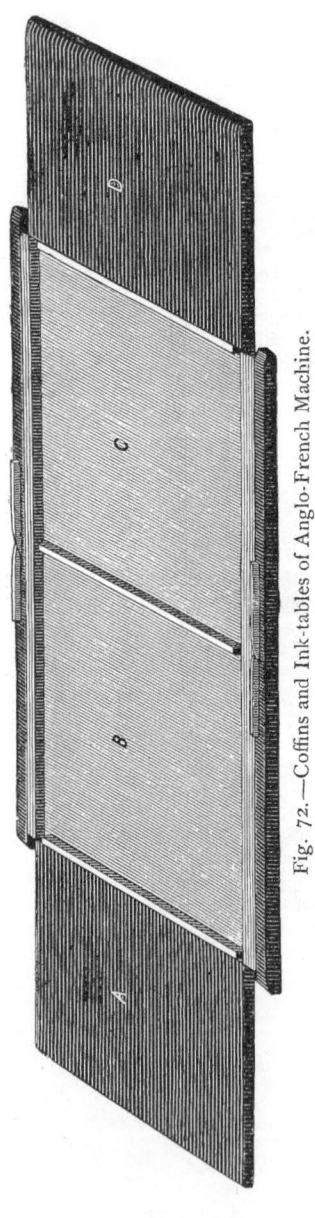

Fig. 72.—Coffins and Ink-tables of Anglo-French Machine.

A Outer-forme ink-table ; B Outer-forme coffin ; C Inner-forme coffin ; D Inner-forme ink-table.

The sheet is held securely by the grippers during the whole of the printing, but tapes are necessary to support, rather than to carry it through the machine, otherwise the extremity would fall loosely upon the forme towards the end of the printing. A short series are fixed immediately underneath (Fig. 74). In the case of the outer they are extended to the delivery. The set-off tapes are carried over a set of guides above the taking-off board, the spindle supporting the same working loosely into brackets on either side. At each extremity are fastened balance weights, which hold the tapes taut, as the tension is necessarily affected by the rise and fall of the cylinder. The pulleys are provided with flanges, and supported by a short gun-metal bracket. At the back is a thumb-screw, by which means the pulley may be adjusted in any position along the bar.

Driving the Ductors.—The ductor roller is driven either by a ratchet or gut band, or by small bevel cogs (Fig. 77).

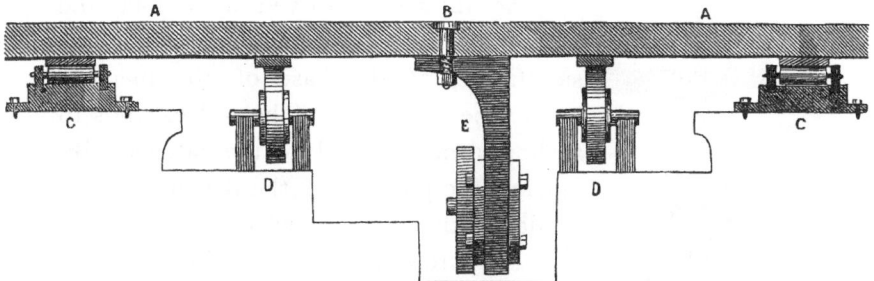

Fig. 73.—Section of Table, showing Impression Pulleys, Rack, and Runners.

A A Tables; B Bolt securing rack; C C Runners; D D Impression pulleys; E Rack.

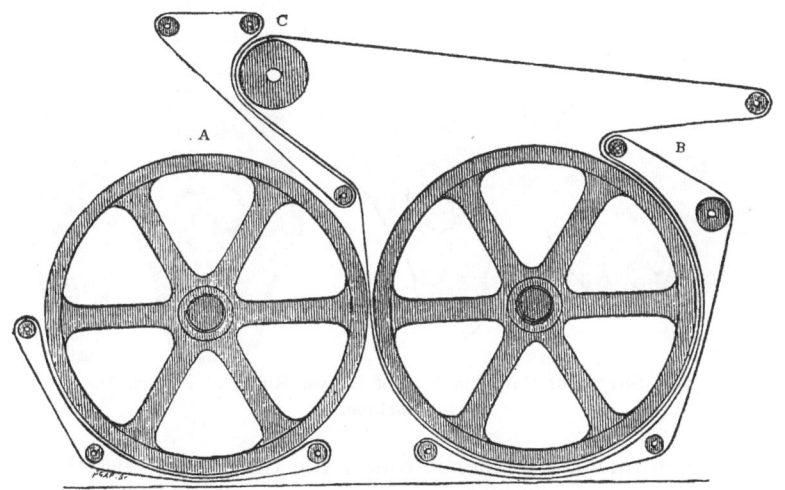

Fig. 74.—Showing Travel of Tapes.

A Entry; B Delivery; C Set-off drum.

The English makers mostly adopt one of the former methods, either of which we think preferable, as in this case the speed of the roller can be adjusted, while with

N 2

the latter the motion is continuous, and the ink must necessarily be regulated by the ductor screws only.

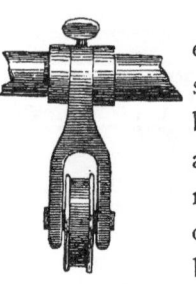

The ratchet is driven by a long rod extending from an eccentric outside the side-frame, at the base of the machine below the cylinders. The rod extends to a short arm attached to the ratchet. By moving the position of the driving rod up or down the arm, to which it is secured by a set-screw, the movement of the duct roller may be regulated to a nicety.

Fig. 75.—Tape Pulley.

When bands are substituted, the end of the duct roller is provided with a graduated grooved wheel, each groove being, of course, of different diameter. The gut travels down to the base of the machine, over two

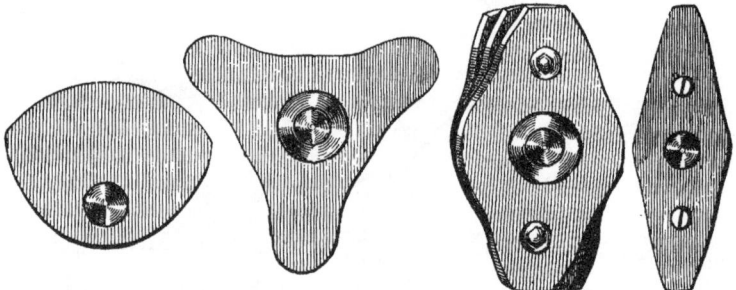

Fig. 76.—Shapes or Cams on End of Ductor Roller of Messrs. Marinoni's Machine.

small runners, and thence round the driving wheel at base of side-frame. The latter is not grooved, as when the band is shifted it may be tightened by readjusting the intermediate runners, which move up and down a small rod, and are secured by set-screws.

When the ductor is driven by gearing, the rod extends from a pinion at the base of the machine, near the driving shaft, to the end of the ductor, the extremity of the

rod being supported in a bearing behind the ductor.
As before stated, we do not prefer this motion, as it
cannot be regu-
lated, and the pre-
sence of the geared
wheels in such a
position, unless
covered in by a
wire network, is
dangerous.

The inking ar-
rangements for the
Anglo-French ma-
chine are excellent.

Fig. 77.—Gearing for driving Ductor.

It will be noticed, on reference to the diagram of the
Napier gripper, that no wavers were employed. The
ductors were placed in close proximity to the cylinders,

and the series of rollers
immediately underneath
both distributed and de-
posited the ink, the dis-
tribution being aided by
a rider which had a
motion from side to side.
The ductors of the
Anglo-French are placed
at each extremity of the
machine in the ordinary
way, and the travel of
the tables is considerably
lengthened. Four wavers

Fig. 78.—Showing Shape on End of Ductor
for working Pulley of Vibrator.

are provided, and the same number of inkers completely
cover the forme.

The Anglo-French Machine with upright Spindle.—The

FIG. 79.—MARINONI'S DUPLEX MACHINE

system of driving the tables in similar manner to that of the ordinary perfecting has been adopted by some English makers, by which means much greater speed can be attained. Powerful cams are employed to raise and lower the cylinder frames, obviating the use of the somewhat complicated rocking frame. With this exception the machine is almost identical in construction with the one just described. It is doubtful if these machines can be considered to have been really successful, possibly because, if speed were required, the ordinary large-cylinder machine was certainly superior. The inking arrangements were of course more perfect, but the feeding-in of single flimsy set-off sheets at a high rate was a constant source of trouble ; and if these were dispensed with, and a single oiled sheet placed round the outer-forme cylinder instead, the set-off apparatus was of no use whatever. The advantages claimed for the adaptation of the upright spindle may be said to be very questionable.

MARINONI'S PERFECTING AND DUPLEX SINGLE-CYLINDER MACHINE.

This machine is in one respect similar to Newsum's, mentioned in the previous chapter, inasmuch as it may be used for perfecting the sheet or for printing two distinct formes.

Although generally identical in construction with the Anglo-French so far as the main movements are concerned, the shapes on the folding frame may be changed, by which means the opening and shutting of the grippers are slightly altered. As will be seen from the illustration, in place of the taking-off board a second laying-on board is substituted, the sheets being carried, after printing, by a series of tapes to the extremity of the machine, down on to the flyers, and duly deposited.

When it is desired to perfect, by altering the gripper shape and adjusting the set-off apparatus the machine is to all intents and purposes converted into the ordinary Anglo-French.

Part II.

——◆◇◆——

CHAPTER X.

MODERN NEWSPAPER PRINTING MACHINES.

The Hoe Multiple-Cylinder—The Bullock—The Walter—The Marinoni Six-feeder—The Prestonian—The Hoe Web—The Marinoni Rotary—The Whitefriars—The Ingram.

THE history of newspaper machines which have passed out of use, and, except in exhibitions or in the lumber-room of some old printing office, are no more to be seen, ends, as we have said, with the *Times* monster vertical machine of 1827.

The first of what may be termed modern newspaper printing machines was certainly the large multiple-cylinder machine erected by Colonel Hoe in 1848 for the Parisian daily paper *La Patrie.* It is needless to say that, during all these years of gradual advance amongst English printers, our inventive cousins on the other side of the Atlantic had not been idle. From the attempts of Dr. Kinsley to turn Nicholson's ideas to practical account, each new discovery had found its way across the Atlantic to be adopted or improved upon by the ingenious mechanics of New York. As early as 1850 we hear of a Hoe machine "not having yet found its way across the Atlantic." Although the principle was identical with the Applegath vertical, already described, the application was altogether different. It was more compact, and singularly free from the complicated

FIG. 80.—HOE'S TEN-FEEDER NEWS MACHINE.

mechanism which characterised that employed by the *Times*.

As will be seen on reference to Fig. 80, the type or plate cylinder was arranged in a *horizontal* instead of a *vertical* position. At intervals around this cylinder were fixed small impression drums, the circumference being slightly larger than the sheet of paper to be printed. Immediately underneath each drum was fitted a series of inkers, &c., which deposited the necessary quantity of ink for one impression. The sheets were stroked down to the mark, and immediately before the forme passed the impression drum, the grippers on the latter seized the paper, and in one revolution an impression was taken—the sheet being released into the tapes and delivered on to the taking-off board in close proximity.

The major portion of the surface of the cylinder served for the inking table. The ink is contained in a large duct placed under the main cylinder, and by means of the ductor roller and vibrator is distributed upon the table.

The original machines were only constructed with four impression cylinders, and with 1,000 from each board would yield 4,000 copies per hour one side, equal to 2,000 perfect. By enlarging the central cylinder and increasing the number of impression drums, it was obvious that the output could be considerably augmented. Eventually, six-, eight-, and ten-feeders were made, and these machines at the time were fully equal to the demand for morning papers, as the circulation was then comparatively small.

It was necessary, however, in order to utilise profitably the eight or ten impression drums, to drive the central cylinder at a great speed. But Messrs. Hoe successfully overcame all difficulties in this respect. The machines were admirably built, and the small amount of trouble they gave, coupled with the splendid results, soon rendered them popular, not to say universal, with newspaper proprietors who were fortunate in having to cater for an increasing demand.

The first Hoe press erected in London was made for *Lloyd's Weekly Newspaper.* Its superiority over the Applegath was soon admitted, and the proprietors of the *Times* ordered two ten-feeders. A condition, however, was imposed—that they should be manufactured in England. Sir Joseph Whitworth was employed, and that firm erected the machines in Playhouse Yard. In consequence, however, of the various difficulties and delays, the inventors established works of their own in London, and subsequently manufactured the presses in this country themselves.

In the original form of this machine the types themselves are, as in the Applegath machine, locked up in "turtles," with wedge-shaped column-rules, which latter are held down to the turtle by tongues projecting at intervals, and sliding in rebated grooves cut crosswise in the face of the turtle. The head- and cross-rules are, of course, curved to the same curve as the turtle, and the whole secured by screws and wedges. This machine, however, was also fitted to receive stereo casts in place of the curved formes, and the application of type to the cylinder is now entirely abandoned in all machines.

Shortly after the Hoe machine came into use, another American, M. S. Beach, applied the same principle to a perfecting machine, placing the second forme upon the other side of the central drum (thereby doing away with the necessity for the balance weights), and turning the sheets as printed to receive the impression of the second forme without stopping the cylinder. Such a machine, with a type drum of only 4 ft. in diameter, and eight impression cylinders, was erected for the *New York Sun*, and is said to have produced as many as 22,000 *perfect* sheets in the hour.

This latter machine never found its way to this country, but in the meantime Middleton constructed a machine upon the same principle as Hoe's, capable of printing 20,000 copies per hour, which was used for printing the *Morning Herald*, the chief difference being the use of one impression cylinder only to every two laying-on boards, there being five of the former to ten of the latter in the machine. The printed sheets were delivered to five taking-off boards in like manner. The ink is supplied, as in the Hoe, by a ductor below the type cylinder, and is distributed in a somewhat similar fashion upon a table attached to the blank portion of this cylinder. An improved distribution of the ink is, moreover, obtained by giving to this table a lateral motion by means of a strap on either side of the machine.

As early as the Exhibition of 1851 Thomas Nelson (of Nelson and Sons, Edinburgh) exhibited a working model of a rotary machine for printing book-work, which is said to have produced excellent results, and to have printed as many as 10,000 copies an hour. The work, however, in the then imperfect state of the machine, was not considered sufficiently good for the purpose for which it was intended (*i.e.*, for book-work), and the machine dropped out of sight for the time being, to be revived again fifteen years later, when the peculiar adaptability of the rotary system to the growing requirements of the newspaper press was perceived by

Mr. Walter, and ultimately by the whole newspaper world.*

Meantime, however, the idea, whether suggested by Nelson's exhibit or not, had been fermenting in the brains of more than one inventor, as an examination of the Patent Office records will show. In 1857 a patent was taken out on behalf of the American machinists, S. W. Woods and H. A. Bills, for certain "improvements in automatic feeding apparatus" for printing machines. In this both the reel and the knife are features of the suggested system, the paper being "laid, in a roll of the required size, on an endless belt of felt, leather, or other material, moving at the same velocity as the types," which latter are to be placed in two series of "many-sided rotating beds," an apparent return to the prismatic drum of Donkin and Bacon. Another suggestion in this specification, since carried out, is the use of two reels, one to take the place of the other without delay in shifting, a system afterwards adopted by Hoe in 1871.

Applegath, too, apparently saw the advantages of the new systems, since in 1859 he patented a web-machine and cutting apparatus, and in 1866, the year in which the *Times* machine first saw the light, this indefatigable inventor is to the fore again with a plan for cutting the sheet *before* damping and printing, the principle of the reel and the knife having been apparently accepted by this time.

* If we are to believe the American press, a mechanician named Thomas Freuck constructed, as early as 1837, a machine to print from a continuous reel, on which he proposed to print a 12mo edition of " Robinson Crusoe." There is no evidence, however, of its ever having actually worked, as the newspapers of the day merely record the fact, without stating whether it ever answered the expectations of the inventor. We give the story for what it is worth.

Although the ten-feeder Hoe was certainly capable of accomplishing extraordinary work when compared with the machines previously employed, the cost for labour was necessarily very great. A ten-feeder machine required about 25 men in constant attendance : 10 feeders, 10 takers-off, men to remove the work, and the printer. So far it was but a slight improvement on the Applegath —except that it was more compact, certain, and simple. However, the ingenious inventors soon applied the flyers, and the cost of labour was materially reduced, the takers-off being dispensed with.

The next practical advance in the construction of the newspaper machine must certainly be claimed by another American—Mr. William Bullock, of Philadelphia, who in 1865 completed *the first really automatic printing machine* It is a very compact piece of mechanism, and although eminently successful, it must be confessed that its adoption was comparatively slow. Whether the printers were satisfied with the Hoe generally used, or were indisposed to incur the further outlay the addition would involve, we are not in a position to say ; but it has appeared to us that the manufacturers have not reaped the reward to which their enterprise clearly entitled them. The fact is, that immediately the automatic or reel principle was demonstrated to be practicable, other ingenious mechanicians turned their attention to further improvements, with the result that, although many Bullock machines almost identical in construction with the original are still being used, others have in some way asserted their superiority, and to a great extent superseded them.

In the following year (1866) the proprietor of the *Times* succeeded in perfecting the Walter press. This like the Bullock, is perfectly automatic, and may be classed among the successful web machines of the present day. For some long time after its erection, considerable secrecy was kept with reference to the details of its construction; but eventually several were made for other papers, and are at the present time employed by the *Daily News* and *New York Times.*

It is perhaps a strange fact that the next newspaper machine which may claim any degree of success was one that required hand-labour. Its introduction can scarcely be attributed to its being superior to the Bullock, but presumably either because it was cheaper, or because printers had not sufficient confidence in the newly introduced automatic principle. We refer to the Marinoni six-feeder perfecting machine. As regards the results the makers claimed, it was an unqualified success. Six men could lay on 1,500 sheets per hour each, giving a total of 9,000—a very large number, especially in the case of the *Echo*, which was printed in duplicate.

The next machine which was introduced to the newspaper world effected a comparative revolution. The Victory Web Printing and Folding Machine was fitted with a cleverly constructed folding machine, by which papers could be delivered, folded, at the rate of about 9,000 copies per hour. The folding apparatus, however, was not its only claim, as the work was well done; and even omitting the folding, it was fully capable of holding its own against those already in use.

A *movable-type* rotary machine, the Prestonian, soon followed. It was made by Messrs. Foster, of Preston. The type was imposed in a curved chase, and was firmly secured by peculiarly constructed steel column-rules, the body being wide at the top and narrow at the bottom. By this machine it was claimed that both time and expense were saved, inasmuch as the forme could be placed on the cylinder and printed therefrom. When we consider, however, that a curved stereo plate could be cast in between ten and fifteen minutes, the gain is questionable, especially as a fount of type would be speedily destroyed, if employed daily for long numbers.

Messrs. Hoe perfected their Web machine in 1873. This compact and beautiful piece of mechanism is justly held to be one of the most successful printing machines in the market. As a piece of engineering skill (so far as the manufacture is concerned), or looking at it as simply a machine by which a large number of papers may be printed in a comparatively short time, with little or no manual labour to assist, it is very highly spoken of by all who employ it. The space it occupies is very limited, the dimensions for a full-sized machine being 20 ft. long, 6 ft. wide, and 7 ft. high. It is capable of turning out about 12,000 full-sized sheets per hour. This machine is employed by several London morning papers.

Messrs. Marinoni introduced in the same year (1873) their Rotary Web Machine, which has also proved to be an unqualified success. It is very dissimilar in its construction to the Hoe, as will be duly noted in the chapters devoted to the respective machines.

We should perhaps mention in this chapter two other rotary machines, which, although in no way claiming to rival those already spoken of, have distinctive peculiarities. The Whitefriars, the invention of Messrs. Pardoe and Davis, was in 1872 erected for Messrs. Bradbury and Evans. Originally built as a fast two-feeder perfecting, the reel was subsequently added, and the machine proved itself capable of producing capital work. It is the smallest of the rotary machines, and, although its capability is not so great as most of those already referred to is well adapted

O

for periodicals, &c., with a large circulation. One has, we believe, been recently built to print *Punch*, and the quality of the cut-work is certainly very good.

In 1876 Messrs. Middleton constructed the Ingram, the invention of Mr. W. J. Ingram and his machine overseer, the late Mr. J. Brister. The feature of this machine was the introduction of a very large plate cylinder, and improved distribution, so as to render possible the printing of engravings. The original is at the present time employed in the production of the *Penny Illustrated Paper*.

Although various improvements and modifications have been made in the minor details of construction in several of the original machines, those mentioned are now generally accepted as being the most successful. At the present time 10,000 to 12,000 copies of a full-sized newspaper may be printed in an hour. It therefore resolves itself into the multiplication of machines to enable any demand to be met. We may mention that the *Standard* and *Telegraph* have twelve and ten machines respectively ; and, notwithstanding the enormous circulation, the requisite number of copies is daily produced between the hours of two and six a.m., with comparatively few workmen.

At the time of writing (1888) the total circulation of the principal daily London newspapers may be roughly stated to be about 1,200,000, and we append the names and number of machines employed :—

Times	Eight Walter Machines.
Daily Telegraph ...	Ten Hoe Machines.
Standard	Six Hoe Machines and Six Prestonians.
Daily Chronicle ...	Four Hoe Machines.
Daily News	Eight Walter Machines, with cutting and folding attachments added by Foster and Sons, Preston.

The Globe	Three Victory Machines.
Echo	Six Marinoni Machines.
St. James's Gazette	One Ingram and one Hoe Machine.
Pall Mall Gazette	Four Marinoni Machines.
Evening News ...	Four Prestonians and one Victory Machine.
Sportsman	Two Victory and two Hoe Machines.
Sporting Life ...	Four Marinoni Machines.

Of the largely circulated weekly papers :—

Lloyd's News ...	Six Hoe Machines.
Weekly Dispatch ... *Referee*	} Six Marinoni Machines.
Weekly Times ...	Three Marinoni Machines.

It must not be supposed, however, that each newspaper uses the whole of the machinery it has at its disposal, as it is obvious that provision must be made for breakdowns. Fortunately accidents are rare, but we think we are justified in stating that every daily paper is provided with machinery capable of turning out 50 per cent. more than the usual demand.

A duplicate set of engines and boilers is also provided, and in fact everything that can be dictated by experience and foresight is to be found in the majority of daily paper establishments.

The only exception that may be taken to the general arrangement of some of the London paper offices applies, perhaps, to the rooms in which the printing is done. Many of the offices are in the most crowded part of the City— in the vicinity of Fleet Street. Space is valuable, and, excepting at an enormous outlay, impossible to obtain ; and the occasional absorption of adjoining premises frequently leads to the adoption of small, dark, low, and unhealthy apartments, where high and light rooms are desirable.

During the last ten years the *Times, Standard, Daily Telegraph,* and *Daily News* have each rebuilt and enlarged their premises, to the undoubted benefit of the workmen. The electric light is being gradually adopted, thus materially reducing the temperature, which, when gas was solely employed, was almost unbearable.

In describing the Rotary Machines, it is our intention to point out the chief peculiarities of each kind. It is obvious that every rotary press must be constructed on the same principle. The web of paper must pass between two different sets of cylinders—the inner- and outer-forme impression and forme cylinders. Every machine, while accomplishing similar results, has its own special features. The disposition of the cylinders may be different; the inking apparatus singular; the manner of taking-off may vary, or the speed be increased; the folding machine may be adopted—but the rotary principle is in all cases worked strictly on the same lines.

It remains to be noted that the ultimate success of the principle involved in its construction was dependent, amongst other things, upon the development of the stereotype process, and its application to the making of curved plates suitable to the type-cylinder. Stereotyping was practised in England, or rather Scotland, as long ago as 1721, by William Ged, but the method of making matrices in plaster-of-paris, from which to cast the types, was only used to produce a flat plate. Cowper, as related in Chapter I., got over this difficulty, in theory at all events, by proposing to curve the plates themselves by passing them between rollers; but this process, though used since his time—it does not appear that Cowper made any use of his own specification in that particular—was at first liable to spoil the plates, and could not be depended upon for accuracy—fatal defects where speed and immediate readiness for use are essential conditions of success, as in the case of newspaper printing.

Even after the general adoption of the type cylinder for newspaper and perfecting machines, the type itself was still used in turtles, necessarily involving much labour, as well as some danger of accident in the locking-up and fixing of the turtles upon the cylinder. Stereotyping seems to have been in the minds of most of the inventors, however. Specification after specification describes the fixing upon the cylinder of the "types, or *stereo-plates*," and Dellagana's discovery, and the use of the *papier-mâché* method, at last made it practicable to cast curved plates by using a flexible matrix. From this time cylindrical stereotype became a practical thing, and the rotary machine had fair play. Marinoni followed the *Times* machine in 1867 with a patent for "cylindrical stereotypes," and although one more attempt was made to print direct from the type (by Duncan and Wilson in 1871), the new method became general in machines of this class. By the process now adopted, the very simplicity of which makes us wonder at its long-delayed introduction, the forme, as soon as locked up, is covered by several sheets of paper, moistened, and beaten with a stiff brush into the interstices of the type. A few minutes suffice for taking the impression, a few more for baking or drying the matrix between hot plates, and a suitable casting-box of the required form receives mould and metal, turning out immediately a curved plate of the exact thickness, which only requires to be trimmed to the necessary dimensions, and planed, before placing it upon the cylinder.

With such an adjunct, as we have already said, the rotary machine passed from the regions of experiment to those of practical utility. The various shapes it assumed will be found discussed in detail in the following chapters, in which we shall endeavour to describe the principal machines in use at present in this country.

CHAPTER XI.

THE WALTER ROTARY MACHINE.

THE *Times* newspaper has always taken the lead in the matter of printing appliances, its natural development continually necessitating the adoption of improved contrivances to meet the requirements of its production. It thus happens that the history of the *Times* newspaper is very closely connected with the history of printing. At first the hand-press, then the perfecting Web, now called the Bar, then the ten-feeder Hoe, and now the Walter rotary!

Apart from the cost of labour on the ten-feeder, which occupied the space of a small house, and required many men to work it, there was the chance of perfecting the sheets the wrong way round (for it only printed one side at a time), and the certainty of getting such imperfect register that it was often impossible to cut open the sheets without cutting off the " heads."

In describing the Walter press we must remind our readers that every part of it was original in its adaptation, the inventors, Messrs. Macdonald and Calverley, having nothing to guide them, unless their minds contained memories of the calico-printing machines.

This machine has a side-frame unlike most others, and seems to straddle, but it is firm, and not unsightly. The reel is at the end, and near the floor, and passes over a roller which tightens it and guides it also. Formerly the paper was hoisted into its place dry, and was damped

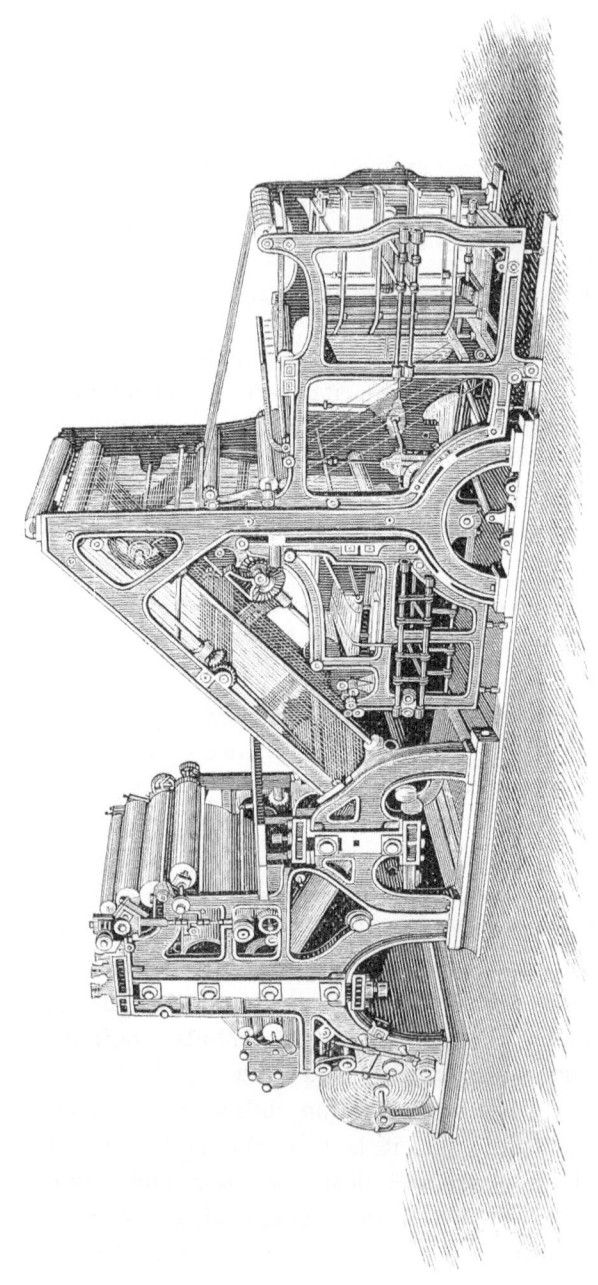

FIG. 81.—THE WALTER ROTARY MACHINE.

in the instant between its unwinding and the first impression. The damping cylinders, now done away with, were hollow, perforated, covered with blanket, and filled with sponge, the water coming in at the ends, and out at the pores. In revolving, this water was thrown out on to the paper, but the printing happened so soon afterwards that the paper could not absorb it and become mellow, as it should do ; so the process, after being carefully and ingeniously contrived, was abandoned, even the other rollers through which the moistened paper used to pass being insufficient to spread the water evenly and thoroughly on both sides of the sheet as it entered the machine. A steam damper was afterwards tried, but with only partial success.

The set-off is dealt with in a very ingenious manner, by a cylinder being placed in contact with the blanket cylinder of the outer forme, which gathers some of the expressed ink, and is itself cleaned by a bar, covered with calico, which touches it at another point from that which comes in contact with the blanket.

This cylinder, travelling so as always to touch the blanket which has the set-off, not only picks up some of the ink, but dries that which remains on the blanket by means of heat, which is produced by friction. Alongside this warm cylinder lies a bar, round which is wound a sheet of calico, and kept in its place by means of a ratchet, so that while the cleaning drum revolves it chafes the stationary calico bar, thereby producing friction, which gives the gentle warmth referred to. When this cloth gets dirty, a short hand-lever enables the attendant to shift it a cog or two, so that a clean part may come against the cleaning cylinder. This is a very effective method of dealing with set-off, and is to be commended, for any adjustment of set-off sheets—or, more properly speaking, endless bands of paper—must have conspicuous disadvantages on a machine which travels so rapidly.

The Walter does not do so well with a flyer as it does with a folding machine attached, and cannot run so fast, because it scatters the sheets, whereas the folder is far neater in its method. The sheets, when printed on both sides, are cut almost off the length, and, entering a race of swifter tapes, are separated completely by the effort thus made by the sheet in getting away, and they are deposited on the taking-off board by the alternate blows of a double flyer.

The printing and impression cylinders are piled up one above the other, the impression cylinders being between the others, and the whole of them being kept from closing down as the bearings wear, by cams on the cylinder shafts, which, by working together, keep all the cylinders at their proper distance. This is a very thoughtful contrivance, much required where the cylinders are piled one above the other, but probably not wanted where they lie side by side.

The doubt arose in the maker's mind whether he should trust to the pull or bite of the pairs of cylinders to conduct the paper safely through the machine, the face of the plates being, after all, only an imperfect cylinder. To assist the paper through, two rollers were contrived, which touched each other exactly their whole length, and, lying one above the other, they wound the paper in with careful tension, being geared by wheels on their shafts to the train of powerful wheels on the off-side of the machine. These are seldom used now, as they have been found unnecessary; but the idea was good, and thoroughly mechanical.

The inking apparatus is well contrived. Rollers of iron and indiarubber and composition are used, and the waving rollers, called distributors, which are not made of iron, can be dispensed with, although it is better to use them. The soft inkers must of course be used, as the iron rollers are only distributors. The rollers in the ink-duct are necessarily of iron, so are the feeding rollers, these latter being turned smaller where ink is not required, just as a composition vibrator is cut away on a book-work machine between the pages, or where the "gutters" come. These, being in contact with the ink-duct rollers, throw a continuous stream of ink on to the other iron distributors, and also on to the rollers of softer material, which is thereby carried to the inkers, and so to the face of the forme. The friction of the iron rollers keeps the ink warm, and also mills it thoroughly before it gets to the inkers, which are large and heavy, and are borne in their places by "carriages" of excellent device and considerable strength.

The plate cylinders are easy of access, the attendant standing just above the reel of paper (which runs on bearings on strong standards resting on the floor) to reach the inner-forme cylinder, and by creeping under the machine, through an arch-shaped opening in the lower part of the side-frame, he can work at the outer-forme cylinder, sitting down to it. A small pit is necessary here to allow room for the man's legs; but the position is not cramped in any way.

When the sheet is printed, such a paper as the *Daily News* can be cut at the heads as well as separated from the reel by the cutting cylinders being arranged to make two cuts to each sheet, the first half being sent along a race of tapes so as to delay its arrival at a point where the second half meets it; the folding machine then deals with it as a whole, and delivers it alternately at two boards, folded and ready for sale. When the sheets reach the "stop" of the folding machine they are prevented creasing—which, going at a high rate of speed, they might do—by being stayed between indiarubber pads, the lower ones stationary, and the upper ones rising so as to admit of the entry of the sheets, and falling so as to hold them momentarily in check. The folding machine is double at its finish, so travels at that part at only half the speed of the printing machine, and the delivery boards are so arranged that they shift at every quire, or other quantity as desired, and so "lay off" the work, and count it with mathematical precision.

The Foster folder does this very satisfactorily, and the two machines thus combined work exceedingly well. The machine, although standing eight and a half feet high, is as rigid as a rock, even at a speed of twelve thousand revolutions an hour, and it is altogether an admirable piece of mechanism. It is 8 ft. long and 6½ ft. wide.

"A few months ago," says Mr. Herbert Spencer, writing in 1872, "the *Times* gave us an account of the last achievement in automatic printing—the Walter press, by which its own immense edition is thrown off in a few hours every morning. Suppose a reader of the description adequately familiar with mechanical details follows what he reads step by step with full comprehension, perhaps making his ideas more definite by going to see the apparatus at work and questioning the attendants. Now he goes away thinking that he understands all about it. Possibly under its aspect as a feat in mechanical engineering he does so. Possibly also, under its biographical aspect, as implying in Mr. Walter, and those who co-operate with him, certain traits, moral and intellectual, he does so. But under its sociological aspect he probably has no notion of its meaning, and does not even suspect that it has a sociological aspect. Yet if he begins to look into the genesis of the thing, he will find that he is but on the threshold of the full explanation. On asking, not what is its proximate, but what is its remote origin, he finds, in the first place, that this automatic printing machine is lineally descended from other automatic printing machines which have undergone successive development, each presupposing others that went before; without cylinder printing machines long previously used and improved there would have been no Walter press. He inquires a step further, and discovers that this last improvement became possible only by the help of *papier-mâché* stereotype, which, first employed for making flat plates, afforded the possibility of making cylindrical plates. And tracing this back, he finds that plaster-of-paris stereotyping came before it, and that there was another process before that. Again, he learns that this highest form of automatic printing, like the many less-developed forms preceding it, depended for its practicability on the introduction of rollers for distributing ink instead of the hand implements used by printers' devils fifty years ago; which rollers, again, could never have been made fit for their present purposes without the discovery of that curious elastic compound out of which they are cast. And then, on tracing the more remote antecedents, he finds an ancestry of hand-printing presses, which, through generations, had been successively improved."

No doubt the chief interest to an ordinary observer of such a machine as the Walter press lies in the history of its gradual development, a history which in the preceding chapters we have endeavoured to sketch more or less fully. Honour, however, to whom honour is due. It must not be forgotten that to the father of the present proprietor of the *Times* belongs the honour of having first applied steam to printing machinery, and that father and son, if they have not invented the various improvements which have changed the Kœnig press of 1814 into the Walter press of to-day, have at least, by diligent research and encouragement, identified themselves with each fresh discovery, and made it in a sense their own.

For the new machine, however, Mr. John Walter did more than that. For four years, in company with Mr. J. C. Macdonald, the manager, and Mr. J. Calverley, the chief engineer of the *Times*, he devoted himself to a series of experiments which culminated in the machine patented in the names of these two gentlemen in 1866. We may properly call the Walter the father of rotary machines, since an examination of the various systems leaves little doubt as to the source from which they derived their inspiration, though each inventor has dealt with the principle as he found it, in his own way.

The illustration shows the form of the Walter press, and it will be seen at a glance that its disadvantage for ordinary printing purposes is its great size—some four times that of the Marinoni, and at least twice that of the Bullock. Its dimensions are—19 ft. long by 6 ft. wide, and 7 ft. high. These considerations of space, however, do not apply to an establishment like that of the *Times*, and in any case the change from the enormous bulk of the vertical machines which preceded it must have made it appear quite small by comparison.

CHAPTER XII.

HOE'S ROTARY MACHINE.

THE Hoe rotary machine stands on a substantial pit, a portion of the largest cylinder being lower than the bottom of the frame, and is a long-shaped machine, different from others, which are either high, or squat, or curved as in the case of the Whitefriars. Unlike other machines, the reel of paper is placed above the side-frames, the paper running downwards, and receiving first the inner-forme impression, and afterwards the outer, previous to being delivered at the end of the machine, printed on both sides, cut to size, and *counted*. Should it be necessary to moisten the paper before printing, this is done on a separate machine, by unwinding it from its original reel on to another reel, the paper passing over a water-spray, and being smoothly and tightly re-wound, ready for hoisting by means of a crane on to its bearings above the inner-forme end of the printing machine.

The paper on leaving the reel, and having been already wetted, is conducted immediately between the first pair of type and impression cylinders, where it is printed by the inner forme, and on issuing thence is taken by the second pair, which, perfecting it, pass it on in their turn to the cutting cylinders, where a serrated knife—the shank embedded longitudinally in one cylinder, and the other receiving the cutting surface in a groove prepared for it—severs the perfect copy from the web; severs it, that is, in all but three small divisions of about one-quarter of an inch, these divisions being left intact merely for the purpose of conducting the following sheet into its place ; they are broken on the sheet entering a set of tapes running at greater speed than the travel of the paper through the machine.

The resistance necessary to give sufficient impression is obtained by the method of adjusting these two cylinders, the ends of the shafts being in bearings, of which there are necessarily two to each cylinder. These bearings are fitted in the side of the solid framework of the machine, and are thus rigid enough to resist the side-thrust which is given every time the plate comes hardest against the blanket on the impression cylinder, as it does at each operation of transferring the ink upon it to the surface of the paper. This is conducted through the machine by the pull which the pairs of cylinders give it while rotating, these receiving their own motion by being geared together, right back to the driving pinion.

The circumference of these cylinders is exactly equal to the length (or width) of the newspaper they have to print ; but the second, or outer-forme impression cylinder, is three times the diameter of the first one, it being so constructed that the set-off may not be so great. As, the second impression is given so soon after the first, this is an important feature, and the machine is calculated to print a great many copies without it being necessary to change the blanket. Should it be deemed advisable, however, to make a change, there are ratchets and rollers within the cylinder by which a blanket twice the required size may be put on before starting. When it has gathered so much ink that a set-off occurs, the soiled portion can be wound up, and the clean part substituted.

The inking arrangements are excellent. The ducts

are at either extreme end of the machine, and the ink is in the usual manner taken by the vibrators, around each of which there are arranged no less than six distributing rollers, four of which have, in addition to their rotary, a transverse motion, which latter is obtained by means of a bell-crank (one to each pair), which is actuated by a connecting-rod from an eccentric on the side shafting.

When the sheet is printed on both sides, and immediately before it is delivered on to the taking-off board, it has to be separated from the length which is following it, which is done in this manner:—Where the division takes place the sheet is held tightly by little buffer-springs, which come just beyond the surface of the cylinder opposed to the knife. The paper having arrived at this point is between two drums, one having a projecting knife running along its entire length, and the other a recess corresponding, into which the knife dips, thereby perforating that part of the sheet which lies taut across the gap. The knife has a serrated edge, and does not cut the sheet thoroughly, as it leaves in three places a piece of paper a quarter of an inch wide, which leads the following sheet on its way. The separated sheets are collected round a drum, and sent down the flyer in quantities of any desired number, and these can again be counted by giving the taking-off board a sidewise or a forward motion, and returning it, alternately, while the printed sheets are accumulating on the board.

When it is desired to fold the work printed on this machine, instead of delivering it flat, by means of the flyer, the sheet, after cutting, runs at an accelerated speed to the first folding drum, and when about a foot of the front of the sheet is round this drum it is held,

P

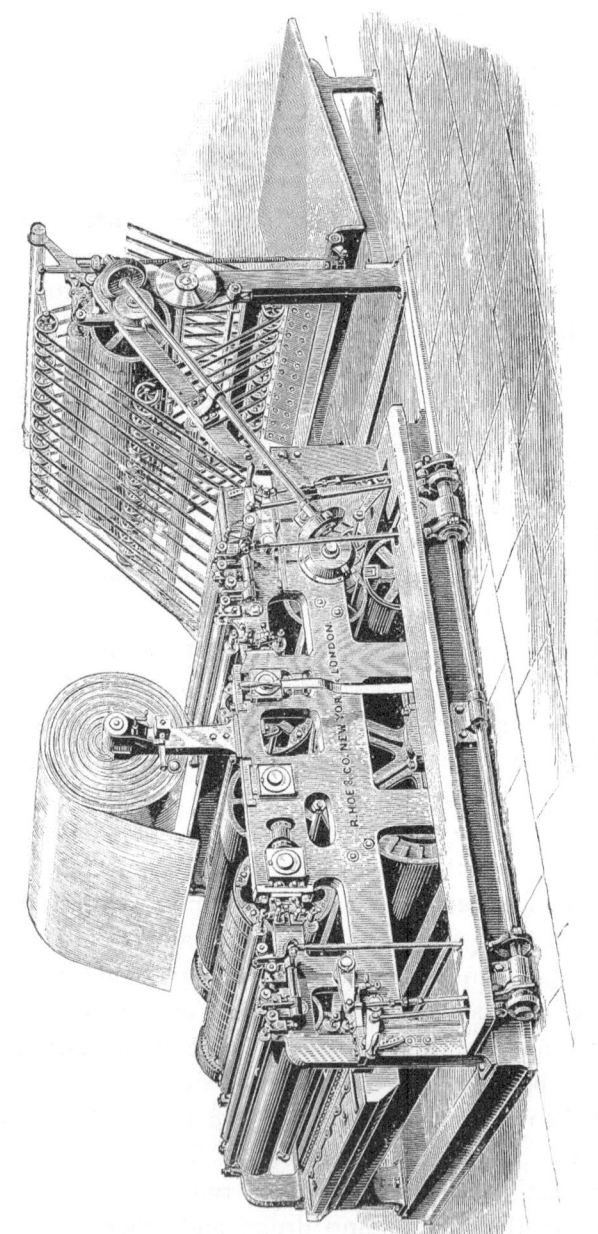

FIG. 82.—HOE'S ROTARY.

to prevent its slipping, until a knife, similarly constructed to the cutting-knife on the printing machine,
but not serrated, strikes it, and thus acting on a sheet
which is lying loose at one end, does not injure it, but
doubles it by this means. The doubled portion is then
taken by a set of grippers, and passed between the second
pair of rollers. This folding machine is peculiar to the
Hoe in some of its details, as when the sheets come first
into the folding machine they are folded in their first and
second folds at a great speed; but as the gripper-drum
cannot be maintained right through at such a high speed,
the twice-folded sheets, as soon as they appear beyond
the second rollers, are divided and sent in two directions,
each alternate sheet being guided by switches into the
different shoots. Thus the folding mechanism which
makes the third and fourth folds, works at only half
the speed of the printing machine, and so allows the
use of folding-blades, sometimes called knives, worked
by the use of simple cranks.

The switches are so placed across the leaving end
of the printing machine as to guide and deflect the
sheets, which, on arriving, are driven through the
folders. The varying speed of the different parts, being
reduced by gearing, is curious and successful too. The
first two folds being given in a way least likely to
break the sheet, enables the other tape-driven parts to
make sure of their work without getting jammed,
which is a great saving of time. When the sheet, already
twice folded, is left by the upper tapes, it is carried
along by the lower set until it reaches a stop, when it is
immediately struck down between another pair of rollers
by the vibrating folding-blade, and is drawn through,
the third fold being then made. This thrice-folded
sheet runs along guides, and passing across another

pair of rollers, another vibrating folding-blade strikes it in the centre, and the fourth fold is made. The four-times folded sheet then glides down in front of a flyer, which pushes it along a sort of table made of five three-quarter round iron rods, from which they are removed, both streams being delivered so close together that one attendant only is required.

Each pair of folding-rollers is so adjusted that they preserve their distance, or give way to the folding-knife, as required, one roller keeping its position always, and the other moving away or closing up by the action of a horizontal spiral spring.

By our description it will be seen that this machine may be worked so as to deliver the printed sheets either flat or folded, and the usual, or most frequent, method of delivery—flat—is greatly assisted by having two pipes, one each side of the flyer, blowing on the number of sheets which come down the flyer at one time, and which would separate, because of their direct downward motion, without it. This wind-blast is worked in a very simple way by means of a small "blower," driven by a gut-band.

The Hoe printing machine has a speed of 10,000 to 12,000 full-sized sheets per hour.

CHAPTER XIII.

THE VICTORY ROTARY.

THIS machine was the invention of Messrs. Duncan and Wilson, of Liverpool, who in 1870 built their first for the *Glasgow Star*. The most characteristic feature of the new machine was the folding apparatus, which, as in the case of the Walter press, was attached to it. Unlike the Walter, however, in which the folding machine is practically separate, and erected at some little distance from the main body of the press, the apparatus in the case of the Victory forms part of the machine itself, and contemporary notices of the invention dwell particularly upon the ease with which the machine turns out its sheets ready folded, and in some cases cut and *pasted*.

The Victory is somewhat elongated, like the Hoe, but differs from it in many other respects. As is the case with all other rotaries, the resistance to the impression is obtained by the rigidity of the frame, wherein the ends of the cylinder shafts revolve in solid bearings. It prints from stereotype plates, and damps the paper the moment before printing, if necessary ; but this plan is not often adopted now.

The roll of paper for the Victory is placed upon low brackets at one end of the machine, and unwound round a small roller just within the frame and slightly below the axis of the large reel itself. The strain thus removed, it returns round the main reel and up to a pair of rollers above the reel which guides it to the top of the frame. In the meanwhile however, it may be damped on both sides by an apparatus peculiar to this machine. Water is forced through a perforated pipe, and the jets issuing from the holes are caught and turned into fine spray by the inter-position of a long, narrow trough. A similar appa-ratus is disposed below the paper on its return to the reel, and a second, placed under the first pair of rollers above the reel, repeats the process on the other side of the web, which is thus as evenly and effectually damped as possible at the speed. Before passing between the printing and impression cylinders, however, it is intro-duced between two upper cylinders heated by steam, the pressure and warmth of which still further distribute the moisture in the paper.

Separate advantages are claimed for every sort of rotary machine, and the position of the cylinders varies, but only to a slight extent ; the folding machine being usually more diverse in character than any other part.

The Victory has a solid frame which carries the entire structure, and in which the heaviest cylinders are fitted, the smaller ones being carried by a lighter frame bolted to it. This machine needs a solidly built pit, and, when fixed in its place, can run at a high speed with good results.

When this machine is built with a flyer so as to deliver the work flat, the frame stands higher than any other part. The sheets run up to the top of it ; the flyer stands erect behind the sheets in their descent, and by the action of an eccentric beats them down on to a table, or taking-off board, from whence they are removed by the attendant. This flyer is made in the ordinary way, of strips of wood called " laths," the thick or bottom end being screwed to a stouter piece of wood, which is, in its turn, bolted to a bar of iron, the ends of

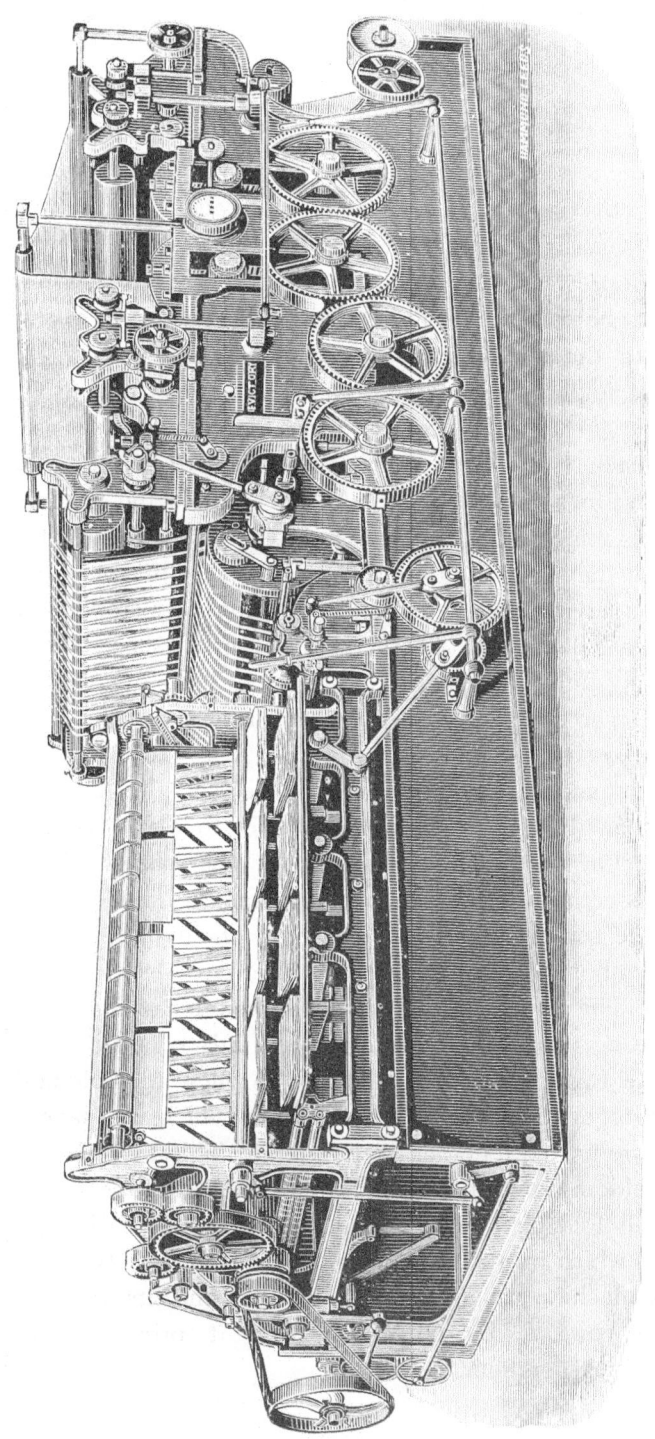

FIG. 83.—THE VICTORY ROTARY, WITH FOLDING MACHINE.

which are held in proper position to act when the sheets are just covering the flyer, when it descends through the quarter of a circle, and returns for another supply, leaving the previous sheets on the table. When the work is required to be folded, the flyer, already described, is taken away, and the folding machine occupies its place.

These folders are various, and may be so arranged as to deliver in four places, all close together, and easily within reach. By subdividing these printed sheets the folders are not crowded, and do not hinder the output of the machine ; and they can be so altered, according to necessity, that half a quire can be folded with one dip ; the sheets having travelled along their tapes to a stop, over two other rollers, will lie there, sheet upon sheet, until the folding-knife above descends, operated on by a cam, and the whole are thrown out accurately counted, it being impossible to make a " miss."

An ingenious adaptation, more used in the United States than here, is sometimes fitted to the Victory machine for the purpose of pasting the sheets of a journal together. A bar supplied along its under-side with paste, somewhat after the fashion of an ink ductor, descends upon the centre of every alternate sheet before cutting. The sheet so pasted is checked for a moment after cutting, and the following sheet falls over it, after which the two pursue their journey through the folding machine together, and the two sheets of which the paper consists are pasted together in the middle. This, as we have said, is seldom done in England, where, indeed, papers are not usually folded so *small* as here described. The Victory machines which print the *Globe* only fold the paper twice.

This machine weighs about ten tons, and its size is 18 ft. 7 in. by 7 ft. by 6 ft. 10 in., and it prints about 10,000 copies per hour.

CHAPTER XIV.

THE MARINONI ROTARY MACHINE.

THE first Marinoni newspaper machine, like the Hoe, was not built on the rotary principle. It was constructed as a six-feeder, but was quicker and more compact than its rival. Originally made for the *Petit Journal*, it was first introduced into this country by Messrs. Cassell, in 1868, to print the *Echo*. It was certainly successful— 1,500 per hour being fed—giving a total of 9,000 copies per hour. Unlike the Hoe, however, the sheets were *perfected*, which gave it a distinct advantage.

The Marinoni, like most other web machines, has been a gradual growth, and the present form of the machine is the result of a series of developments which have greatly altered its form.

In 1872 M. Marinoni took out a patent here (No. 2,041) for several improvements in printing machinery, comprising the application of the continuous web, combined with improved methods of cutting and delivery.

The Marinoni Rotary Machine consists of six distinct parts, which are: 1. The feeding apparatus, including the appliances for supporting, &c., the reel of paper; 2. The wetting apparatus, consisting of the water-trough and the water-distributing cylinders; 3. The cutting apparatus, consisting of the cutting cylinders and the serrated knife; 4. The inking apparatus, consisting of the trough, ductor, reciprocating roller, distributing cylinder, and inking rollers; 5. The printing apparatus, consisting of the impression cylinders and the plate cylinders; and 6. The

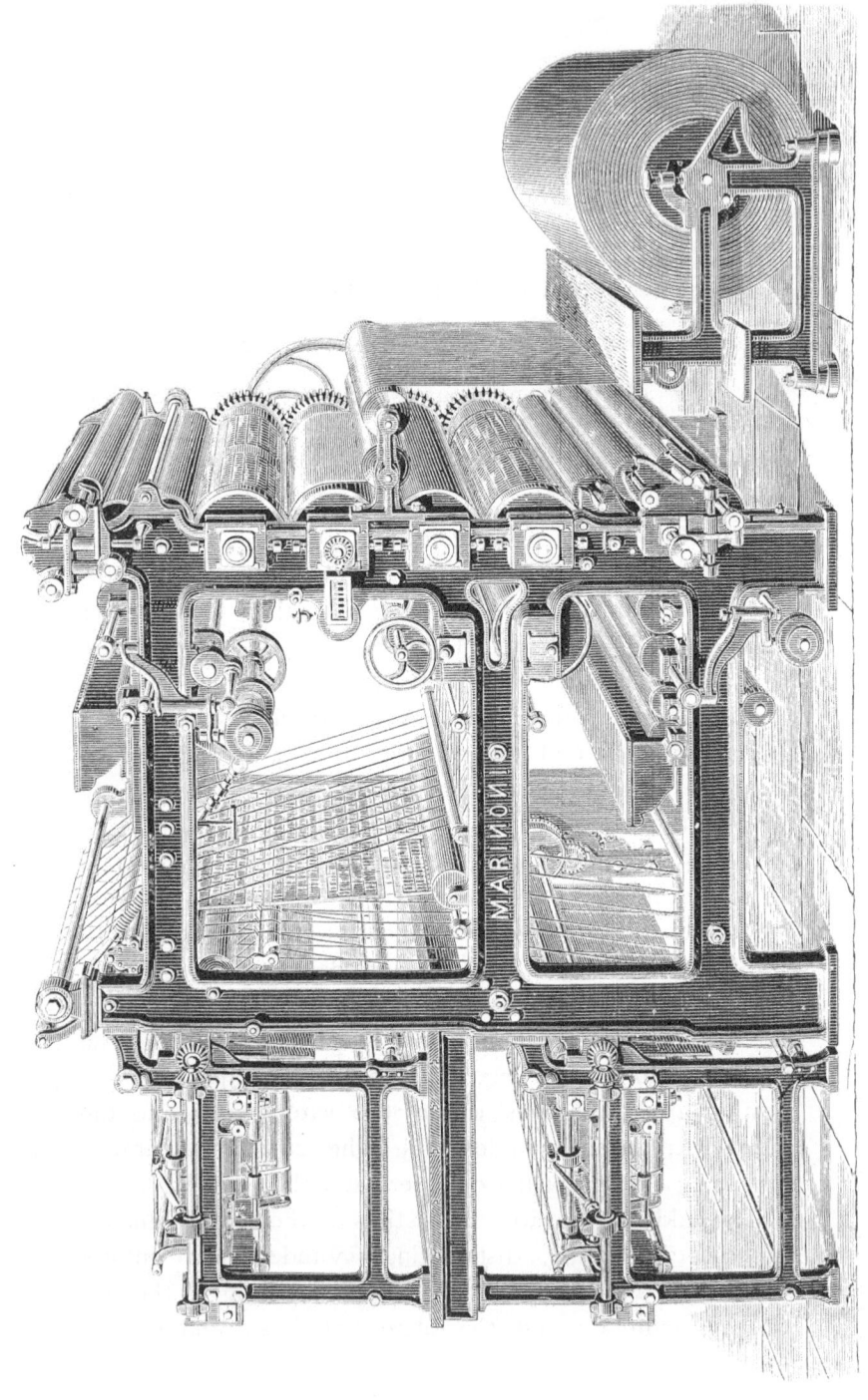

FIG. 84.—THE MARINONI ROTARY MACHINE.

delivering apparatus, consisting of the mechanical arrangements for dividing the streams of sheets, (*a*) into lateral streams, (*b*) into an upper and a lower delivery at the two sides of the machine respectively, the flyers, and the four delivery boards. The paper is conveyed from one portion of the machine to another by carrier cylinders or drums, and by tapes. The accompanying diagram will serve to show the situation and relations of these different parts, representing, as it does, the newest form of Marinoni machine, embodying the latest improvements. The most notable of the improvements, perhaps, is the arrangement patented by Mr. Sauvée, the London manager of the firm, in 1880, by which a forme of any size can be printed, the sheets being cut before entering the printing apparatus, by a knife adjusted to suit the size required.

The chief features which distinguish the Marinoni press from other machines of the web type, are the feeding, the cutting of the sheets before instead of after printing, and, lastly, the delivery apparatus. On some machines the feeding apparatus differs from that in use in most other machines by being double. Two rolls of paper are provided, being placed somewhat differently from the usual method—high up instead of near the ground. They are raised into this position by hoists, and the end of the roll not in use is fixed in a clip. When one is exhausted the other is attached without stopping, while the empty roll is removed to make place for a fresh supply. In order to arrange for this, it has been necessary to duplicate also the wetting apparatus, which consists of a trough filled with water on each side of the machine, into which a small cylinder dips, the paper passing over a second cylinder or drum in connection with the latter.

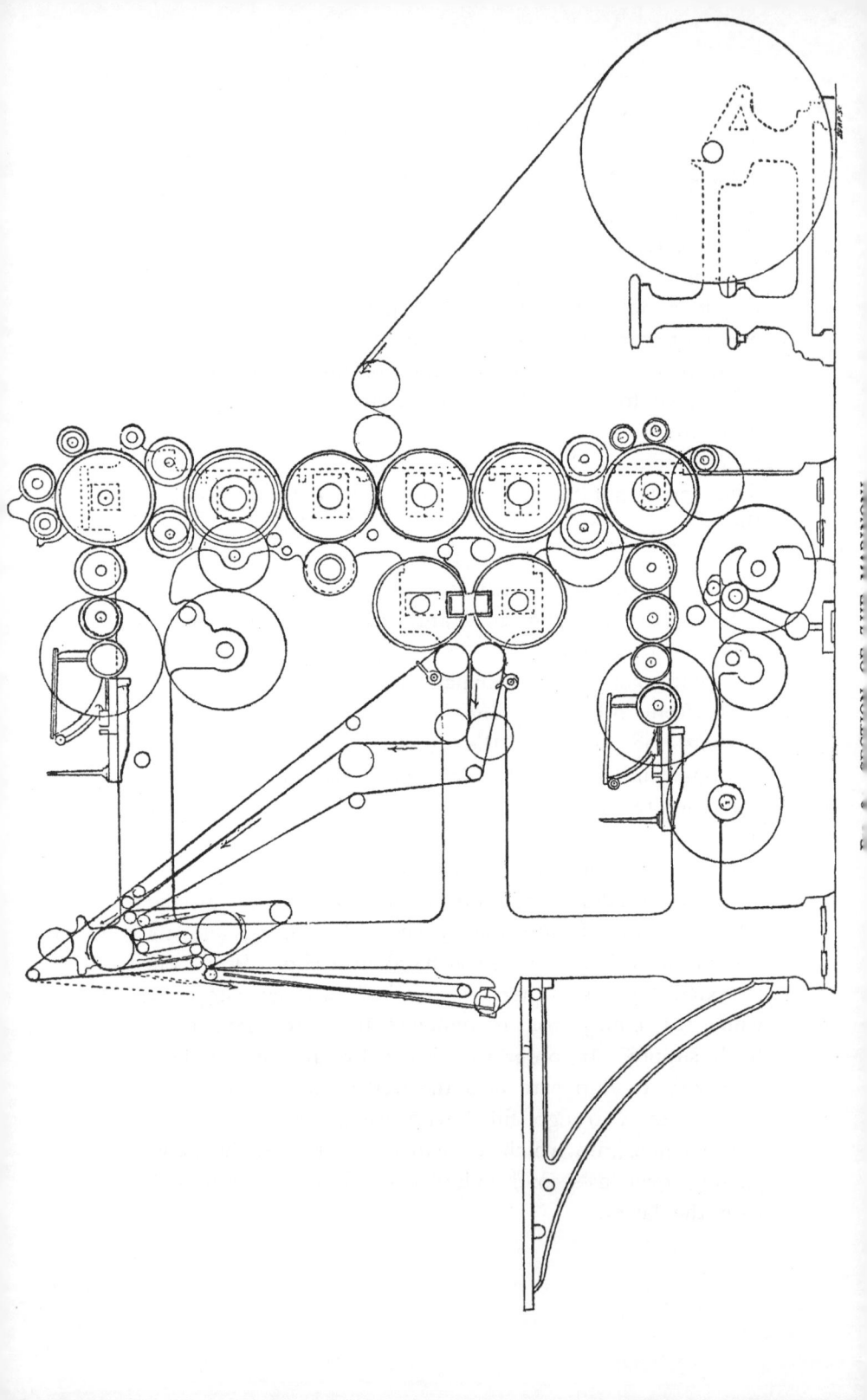

Fig. 2. SECTION OF THE MACHINE

The cutting of the paper before printing differs from the process in use in almost all other machines. In the original form of the Bullock this plan was adopted, but Kellberg's apparatus applied to these machines is said to have increased their speed and general efficiency.

The principal difference, however, to be noticed in favour of the Marinoni machine is the very excellent delivery apparatus. The Bullock press has but one delivery, and while the Walter has two, the advantage is somewhat counterbalanced by the necessity of providing an attendant to receive the sheets in each place. The Marinoni, however, has no less than four delivery boards, while the sheets are delivered so evenly that they require no attention other than the removing of them from time to time as they accumulate. The additional speed gained can be easily understood, while the separation of the sheets makes the application of folding machines, when required, of greater simplicity, the difficulty generally being that a folding machine cannot be run at the same speed as the printing machine itself.

Add to this the accumulator, a feature of the newest machines, by which the sheets are collected in quires, or any number desired, and a delivery as nearly perfect as can be conceived may be fairly claimed for the Marinoni press ; while the almost automatic nature of the machine may be understood from the statement made by the manufacturer, that one man can with ease mind two machines.

CHAPTER XV.

THE FOSTER ROTARY WEB (TYPE, STEREO, AND BILL) MACHINES.

THE idea of printing from type instead of stereotype plates, to save time, may be a relic of bygone fashion. This was accomplished on the Hoe ten-feeder. But as we have before mentioned, stereotyping is now so rapidly done that little if any time is lost in the process, the different pages being dealt with by separate sets of hands.

Foster's original Web Perfecting Printing Machine prints from type, but it may now be relegated to the past. With the type web there is no difficulty in fixing the formes, because the type cylinder is seven-eighths and a sixteenth of an inch smaller than the printing line, the length of the type making up the difference, the "turtles," or galleys, in which the type is fixed being considered part of the cylinder. The sheets are turned out printed on both sides at one operation, as with all recent rotaries, the register being exact, the principal wheels having all the teeth machine-cut.

The folding machine which is added to this machine is on the plan of the Livesey, and directs the sheet, when it is separated from the web, along tapes to its proper position, with complete accuracy; and four folds, if necessary, can be made by means of the guiding tapes passing to and over as many pairs of rollers, fitted with the familiar vibrating folding-knife to strike it through, and send it on its way to the vibrating flyer at the finish. Impression is obtained by contact between the face of the type and the printing side of the blanket cylinder, there being in this, as in the other rotaries, no "bearers" at the sides of the formes to regulate the touch. This important omission compels accuracy of gearing, as any stoppage in the run, caused by unequal teeth, or the slip, or "chattering," called "backlash," wears the type or plates, and, by thickening or blunting the face, spoils the appearance of the work.

This type-printing web machine is made to print from stereotype plates by making the stereo-beds as much thicker than the type-beds as will equal the difference in height between the movable type and the stereo plates. Stereo pages may be used on this machine along with pages of type, which is sometimes found to be a convenience.

The massive frame of this machine carries all the parts above the floor-line, and a pit is not absolutely necessary, but is always desirable, for convenience of oiling and repairs.

The speed is 12,000 an hour for an eight-paged newspaper similar to *The Times*, with open delivery, and only a trifle less with the folder attached. This machine is 20 ft. long, 7 ft. 6 in. broad, and 7 ft. 6 in. high.

The Prestonian, a somewhat similar machine, was the forerunner of this one—the Foster—this latter being improved in many particulars.

THE FOSTER ROTARY STEREO MACHINE.

Another machine, for stereotype only, bears the familiar name of Foster, it having been built to meet greater demands than the type-printing machine could accomplish. With the former class, where the type

was used, only one machine could be employed, but
with the additional help of stereotype, plates can be
duplicated, or any number can be made.

This is a long-shaped machine, with the principal
cylinders placed in the centre, adjusted in massive
bearings in a rigid frame, their general position being
in the form of a diamond. Two of these are the plate
cylinders, and the other two are the blanket cylinders,
reciprocating their impression in pairs. In passing
between the first two cylinders, one side of the sheet
receives its impression, and the second pair gives impression to the other side, when the sheet, still part of
the web, passes on to the cutting cylinders, and is
separated. The separated sheet is at once captured by
tapes, travelling at a higher speed than the machine,
and, an eight-paged newspaper being printed, the first
four-paged sheet enters the race of tapes, from which it
emerges in exact time to meet the second four-paged
sheet, and the two then proceed to the folding machine,
where they receive three folds by means of as many
pairs of rollers and vibrating blades, and are ready for
the hands of the public, folded neatly to size, and the
heads cut. From the time of entry to the time of completion is about two seconds for each newspaper; while
the papers follow each other at intervals of only a
quarter of a second, or, in the case of four-paged papers,
in half that time.

The arrangements for ink supply are necessarily
similar to those of other machines of this class, but are
elaborated somewhat. Two revolving drums (answering
to the ink-slabs of flat machines) distribute the ink
accurately to an even film.

The reel of paper is placed shoulder-high at one
end of the machine, and the folder is at the other, the

cylinders, rollers, ink-ducts, and revolving ink-drums being centrally placed ; the whole machine being compact, rigid, and light to run.

FOSTER'S ROTARY BILL MACHINE.

As newspaper contents bills are wanted quickly, and in considerable quantities, Messrs. Joseph Foster and Sons have produced a rotary machine specially constructed for the purpose. By its use the latest item of news can be inserted in the contents bill.

This machine prints from type, any size from "four-line" upwards, it being cut on the curve, so that it makes up the required diameter of the cylinder when it is in its place. Any fixed length of bill may be printed, and the width may vary. The reel of paper is fixed near the floor at the front of the machine ; next above it is the impression cylinder, and above that is the type cylinder, between which two the sheet enters, and is printed, of course, on one side only. It is then cut from the web, and taken to two boards at the back, where the sheets lie until they are removed by hand.

The type (curved to the radius of the cylinder, which is breast-high) is put in its place as on a galley, and secured by wedge-shaped "white lines," which act as the keystone of an arch does, and keep the whole together. The bill is "two-set," so that it prints all round the cylinder.

This machine is built strictly on the rotary principle, has considerable strength, and runs at the rate of 14,000 copies an hour.

Q

CHAPTER XVI.

THE BULLOCK PRESS—THE NORTHUMBRIAN PRESS— FARMER'S WEB PRINTING MACHINE.

MR. WILLIAM BULLOCK, of Philadelphia, justly claimed for the rotary press which he completed in 1865 the honour of being the first really automatic printing machine. Although the use of it has been almost entirely superseded in London by more recent improvements, the fact that several are working at the present time justifies us in describing it among modern newspaper machines.

The wetting is done (before placing the reel upon the machine) by means of a fine spray, through which the paper is passed before being mounted into position. It is supported on brackets above the machine, and the first process is the cutting of the sheets by a serrated knife rising and falling on a cylinder, but this distinction was removed in some machines by a modification introduced by Kellberg, who placed the knife in its usual position, *following* the printing cylinders.

The paper having been cut, the sheet is seized by grippers on the slotted cylinder which receives the knife, and taken by the impression cylinder, receiving its first impression from the inner-forme cylinder. It is then seized by grippers on the large drum, and carried beneath the outer-forme cylinder, thus perfecting the sheet and leaving it to be carried round by the large drum to a point near the base of the machine, where it is delivered to endless belts, the grippers on which

carry it above the delivery board. A set of flyers working between these belts is worked by a rod in connection with a cam on the distributing cylinder, and, acting
intermittently, detaches the sheet from the grippers and
strikes it down upon the board, the grippers themselves
being opened and shut by cams placed within the circuit
of the belts.

The inking apparatus differs very little from other
machines of the class. Two vibrators take the ink from
the ductors, which are placed under the distributing
cylinders on either side of the machine. Four wavers,
having a side motion, are placed on each cylinder to
distribute the ink evenly over its surface, while the
inking rollers are situated between, and in direct contact with, both the distributing and type cylinders, thus
causing a steady and continuous supply of ink to the
latter.

The Bullock machine is now but little used for
newspaper printing, though it is still at work in the
Queen offices, and for many years printed both that
paper and the *Field.* Its advantages, as claimed, con-
sisted mainly in the comparative simplicity obtained by
the use of grippers in place of tapes, and economy in
space over other machines of the class, 11 ft. by 6 ft.
being the amount of floor-space required for a Bullock of
the *Times* size. This last consideration, however, is no
longer of the same importance as it was in the days of
small machine rooms, while the advantages gained by
the abolition of the tapes are more than counterbalanced
by the difficulty of keeping in order the large number
of sets of grippers (seven in all) which take their place.
These reasons, and the rapid improvements in other
quarters, have probably led to its virtual abandonment
in favour of more modern machines.

Q 2

THE NORTHUMBRIAN PRESS.

Another machine which, though it is not now to be seen in London, is probably still to be found in the north of England, is the Northumbrian, the invention of Messrs. Donison, printers' engineers, of Newcastle-on-Tyne. Its history was summed up to the writer by a leading manufacturer in a somewhat Irish fashion : " There was never but one built," he said, " and now they have all disappeared." This statement, however, was certainly incorrect, since the *Edinburgh Evening News*, the *Sunderland Daily Post*, and the *Dundee Evening News*, with other of the northern journals, were, and possibly may be still, printed upon this machine. It differs little from others already described, being, however, of the simplest pattern, the impression cylinders being placed horizontally in a frame, the two impression cylinders in the centre, and the distributing drums on the outside, the type cylinders working between the two, while the paper is placed in two rolls, one at either end of the machine, on standards, and so arranged that the sub-stitution of one roll for the other occupies but a few moments only. The paper is printed in the web, and cut, after passing through the machine, by cutting cylinders placed below. It is thence conducted, by means of an oscillating frame, into sets of delivery tapes on either side alternately, being thrown thence by flyers on to the delivery boards.

Almost the only improvement worthy of notice in this machine was the placing of an additional roller to assist the unwinding of the reel. This roller is placed directly above the reel, and works upon the inner-forme cylinder, reducing by its action much of the drag upon the impression cylinders.

FARMERS WEB PRINTING MACHINES.

Another machine deserves a few words in this chapter. The inventor, Mr. James Farmer, was a calico printer, of Salford, whose experience, like that of Nicholson, led him to some ideas on the rotary system as applied to printing machinery. With this idea he built an enormous machine, 32 ft. in length, the peculiarity of which was its adaptation to print from two reels placed one above another.

The Farmer has no underneath gearing nor cylinders, and one of its peculiarities is that it requires two reels of paper, by which means its output is increased, although only one set of stereotype plates is required. The standard which holds the reels is separate from the printing machine.

There are three distinct pieces to this machine—the paper standard, a tall upright as high as the printing portion ; the cutter, and the gatherer and folder. The paper runs into the machine at the top, and, descending between the relative pairs of cylinders, is printed in somewhat the usual way, after which it crosses in tapes to the cutter, and is piled up in quires, or any regulated number, by the automatic action of the board, and, after cutting, collected in parcels of a dozen, when a blunt knife descends, and folding them in half, delivers them into a trough.

The whole arrangement of cylinders, rollers, and ink-ducts is supported in a frame of triangular shape, standing on one of its sides ; the blanket cylinders are twice the size of the plate cylinders ; and the ink-ducts have notches in them, through which a sufficient quantity of ink is supplied. The inkers are adjustable at either end by means of thumb-screws, which regulate their pressure on the forme, an item of considerable importance if the rollers get soft in the summer or shrink in the winter.

The damping apparatus is peculiar, consisting of two soft indiarubber cylinders constantly supplied with moisture.

Very few of these machines are now made.

247

CHAPTER XVII.

THE WHARFEDALE ROTARY MACHINE—THE WHITE-FRIARS—THE JULES DERRIEY ROTARY—A ROTARY HAND-PRESS.

THE Wharfedale rotary manufactured by Messrs. William Dawson and Sons, of Otley, is very compactly arranged. The reel of paper is laid at the inner-forme end of the machine, on centres, so that the tug of the cylinders pulls it round easily and without breaking the sheet. Standards, with free rollers fitted at the top extremity, carry the unwinding length of paper to a point where it can descend directly between the first pair of cylinders, where the inner is printed, and immediately runs between the second pair of cylinders, perfected, and is then cut from its length. The sheet is immediately directed into the folding machine, which is attached close to its outer forme, and is speedily delivered, neatly folded, in a position which is convenient for removal.

The side-frames of this machine are heavy, and necessarily very strong, as on them depends the rigidity of the whole structure, the cylinders being carried at their ends by strong bearings set in the framework, and easily accessible. The first pair of cylinders is above the second set, and a little nearer the reel, and the partly printed sheet has to descend to the second pair, after which, being cut, it rises to the highest point of the folding machine. This machine works from stereotype plates, which, with the cylinders, are the exact size of the sheet to be printed. The inking arrangements are

FIG. 86.—THE WHARFEDALE ROTARY MACHINE.

well considered and effective. The ink is well distributed before it comes into contact with the face of the revolving plates, and the ducts are so constructed that the movement of a single screw will alter the flow of ink on to the vibrator. The motion is imparted to the ink-ducts by means of a rotating crank-pin on a bar near the centre of the machine, low down, which controls two levers, one to each duct, the lower ends of both being attached to the pin, as already stated. The chopping motion of the folding blades is caused by cock's-comb segments, which are so placed as to be within view while running, and are worked in a similar way to the ink-ducts. This machine does not require a pit its whole length, but it is necessary to have a small one in the centre, as one wheel of the outer-forme pair lies below the bottom line of the side-frame. Its speed is 10,000 revolutions an hour.

THE WHITEFRIARS.

The Whitefriars is unlike other rotaries in several important details. In the first place, its shape is different, its frame being like a horse's shoe standing upright. Its principal cylinders are grouped somewhat after the manner of an ordinary Bar machine, the inner- and outer-forme cylinders being where in the Bar machine the impression cylinders are, and the impression cylinders of the Whitefriars taking the place of the turning and register drums. The inking cylinders are outside at either end, the ink-ducts being beyond them.

This machine was originally intended to be a fast perfecting two-feeder, but it has since been worked from the web. There are two laying-on boards, one at each end, over the machine, and not far apart, the "striker-board" being placed so that the layer-on can easily strike off, and skid the hand-wheel if necessary. The

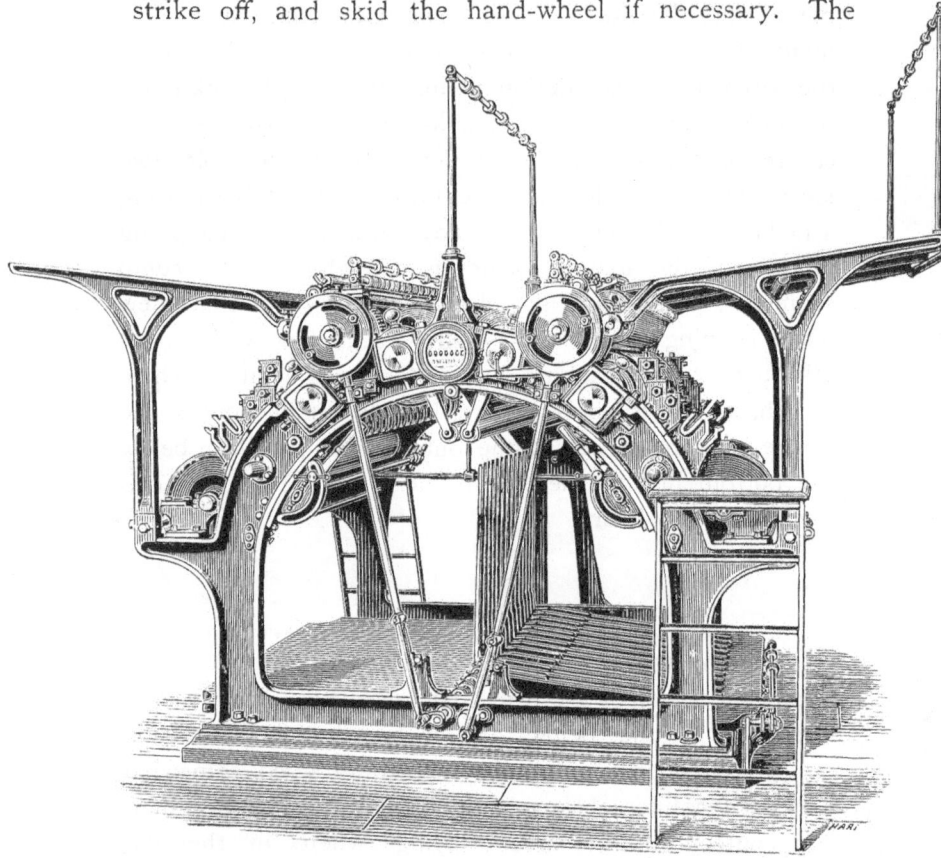

Fig. 87.—The Whitefriars.

driving gear is at the off-side of the machine, supported on an **A** carriage about shoulder-high, and the driving and loose riggers are capable of accommodating a driving belt 5 in. wide, the value of which is often overlooked.

The cylinders intended to receive the plates are curved spirally, with grooves cut and under-cut, so as to enable the screw-catches to travel to any part of the cylinder, and hold the plate securely ; the size of the plate being of no consequence, as the grooves are so arranged that any plate half the circumference of the cylinder and a couple of inches short of its length may be printed, or, in fact, the smallest octavo. The plates may be a pica (the sixth of an inch) or a quarter of an inch thick, the latter for preference, or even thicker. The plates are cast flat, and curved afterwards.

The page being sent into the foundry from the composing room, it is dealt with in the ordinary way by paper process, and when it is finished, bevels cut, and all the trimming done, it is laid on the hot chamber, near the metal pot, and allowed to lie there until it is soft. It is then placed in a bending-box, where it receives its proper curvature, which enables it to lie on the plate cylinder.

The bending-box is made of iron, heated inside by several jets of gas, and it is the size and shape of the printing cylinder— if anything, a little sharper in its curve, because the metal, in cooling, is liable to contract. Above this section of a cylinder is a curved cover, made to fit to a nicety, which is wound down by means of a screw and a hand-wheel, when the flat plate is in its place on the box which is to give its under-side its proper shape. On the face of the softened plate a piece of blanket is placed, and then a spring, made of steel plates, on top ; the side gauge (a straight piece of iron, with a wooden handle, and resting in two slots) is then removed, and the cover wound down. The plate lies under this pressure for about a minute, or less ; the cover is wound up, the spring removed, together with the blanket, and the plate appears curved backwards. This is then taken away, and laid on a cooling saddle to set firm.

This is a section something similar to a "turtle," and allows the plate to cool in its proper shape, without distorting. The plate is then ready for the machine.

Arrived in the machine room, the plate is laid on the cylinder, and screwed up : the impression is tried, and found good, unless the blanket on the impression cylinder is old, or the metal of the plate has cooled too quickly. The rollers are put in their brackets and made fast, paper is got ready on both boards, the smooth side *up* on one board, and *down* on the other, and all is ready for a start.

The sheets follow each other into the machine alternately, and at the rate of about 2,000 an hour from each board. The sheet travels through the tapes until it is printed on both sides ; but its speed is accelerated then, and it glides out, supported by carrying tapes, and when out its full length is beaten down by a flyer, which, hinged near the base of the machine under the inner forme, extends the points of its laths across the board which lies on the bed-plate, and knocks the printed sheets into a pile. This beater is worked by an eccentric and a lever, attached to the shaft of the inner-forme plate cylinder. Some of these machines have a double flyer, but the printing portions remain the same.

If it be desirable to print from the web, an additional structure has to be put up, which unwinds the paper, and, *cutting it before it enters the machine*, sends it through faster tapes, one sheet following the other with an interval of two or three inches, according to the size of the sheet. On this plan that which was just described as the "strikerboard" is thrown out of use, all the sheets going in from one point.

When separate sheets are used, they enter the machine by means of the drop-bar.

The machine we have here described measures 10 ft. long, 6 ft. 6 in. high, and is 8 ft. wide.   The plate cylinders are 5 ft. long, and nearly 4 ft. in circumference, and at each end may be placed three inking rollers, all in contact with the forme, one intermediate roller, two distributors, and one feeder, or vibrating roller.   The ink-drums have a sideway motion, as well as a rotary one, and the distribution is consequently good.

The machine is massively built, the gear-work is all cut, and wide, and the shafts of the four principal cylinders, which have to bear all the thrust of the impression, are 4 in. thick.   This machine does not require a pit, and would be no better for having one, as all the working parts are above ground and well in sight.   It was the invention of Mr. Joseph Pardoe, of Messrs. Bradbury's.

THE JULES DERRIEY ROTARY.

This French machine is somewhat crowded in the plan of its construction, and, to our mind, has too many wheels in gear.   Its chances of chattering, delay, and backlash must be considerable, although one of these machines has for years, until recently, been employed to print a London newspaper, and has done good service.

M. Derriey has an improved machine, working with open delivery or folding machine attached, printing five, six, seven, eight, or nine columns, delivering the papers by quires, folded or flat, and in four-page or eight-page editions, the eight-page editions being made up of two four-page sheets, superposed, pasted in the margin, and folded, all the pages being the size of the London *Times.* Dimensions, 11 ft. long by 7 ft. wide.   Highest speed, 12,000 an hour.

Another machine by this eminent firm is an all-size rotary for jobbing work, feeding either from the reel or by hand, and delivering the sheets either flat or folded. It is so constructed that the overlays may be put under the blanket as easily as on a Wharfedale single-cylinder.

One more among many is a rotary for illustrated newspapers or book-work, the engravings *being worked separate from the text,* with special inking for the cuts. The delivery is by tapes so arranged as not to mark the face of the work, no matter how "full" the cuts may be; and as both impressions come on one side of the sheet, the machine may be used for two-colour work.

A ROTARY HAND-PRESS.

As this interesting piece of mechanism is probably not so well known as it deserves to be, we give a brief description of it, accompanied by an illustration, which may be found in an old volume which was published in 1825 by Messrs. Knight and Lacey, of Paternoster Row, London. This was the first attempt at rotary printing, and it is therefore curious.

The type was fitted on a square arrangement, so mounted on its frame that it could be easily turned, another unequally shaped roller being placed beneath it, and the sheet of paper was conducted between them. The type was inked by means of a roller placed above the type carriage. Leather, stuffed in the manner of a cushion, was first used, but did not succeed, because it became indented with the type; but afterwards a glue-and-treacle roller was used. As this hand-machine is now entirely obsolete, this brief notice will suffice, but the gear-work is worth looking at, it being shown in the accompanying sketch in detail at the top right-hand corner.

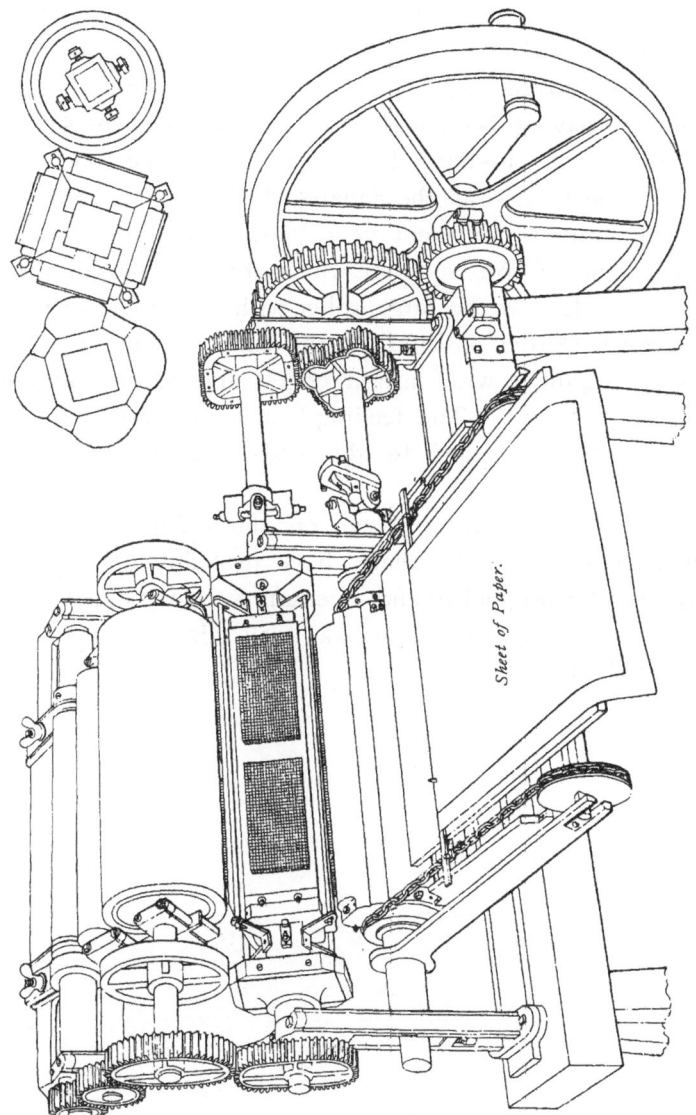

Sheet of Paper.

Fig. 88.—ROTARY HAND-PRESS.

The square type carriage has a square cog-wheel, slightly rounded at the corners. The impression cylinder, if we may so call it, has a somewhat star-shaped wheel, provision having been made for the corners of the square wheel, and the inking roller has a round wheel because it was free to rise as the square wheel drove it upwards, it being only in sufficient gear to turn one intermediate roller, made of soft material, which touched the ink-duct with one side, and the large roller with the other, keeping in its own geared position all the time. The type was placed in galleys with mitred edges, and secured at the ends with screws. The feeding-board was worked by a spiked wheel attached to the ends of the impression cylinder.

The curious gearing shown on the off-side, and the universal joints on the shafts, have probably never been surpassed for originality and ingenuity. It was announced that this quaint contrivance was well adapted to do the best kind of work.

CHAPTER XVIII.

THE *ILLUSTRATED LONDON NEWS* AND THE INGRAM
MACHINE—ANGLO-AMERICAN ROTARY—ON STEREO-
TYPING FOR ROTARY MACHINES.

IN 1876 Mr. W. J. Ingram, assisted by the late Mr. James
Brister, introduced the Ingram rotary. It was specially
designed to print cut-formes of the *Illustrated London
News* at a great speed, and, while reducing the cost of
production, would allow the formes to be sent to press
at least a day later than previously. The method and
objects of the patent are clearly enough set out in the
specification.

"It has been found in practice," says the patentee,
"that 'cuts' or engravings require much more careful
inking than the letterpress, and that the ordinary inking arrangements, which are found to answer very well
for printing letterpress, will give but very imperfect
work from engravings or 'cuts.' It has also been well-
nigh impossible to obtain satisfactory impressions from
'cuts' or engraved plates bent to the sharp curve required to correspond to printing cylinders of the ordinary
size.

"In order to overcome these difficulties I considerably increase the diameter of the printing cylinder to
which the 'cuts' or engraved plates are to be adapted,
so that the curves to which these 'cuts' or engraved
plates are bent may be gentler and of longer radius
than the curved surface of the other printing cylinder.
By this means I am also enabled to place on the same

R

printing cylinder two, three, or more copies of the 'cuts' or engravings, so that while the surface speed of the large and small printing cylinders is the same, the small cylinder, if it contains only one set of stereotype plates for the letterpress, will rotate two, three, or more times for every revolution of the large cylinder. The impression cylinder, which acts in conjunction with the large printing cylinder, is also correspondingly increased in size, and rotates at the same surface speed. If desired, the type cylinder may be increased in size so as to be capable of receiving a duplicate set of stereotype plates for the letterpress, while the large cylinder will have a triplicate or other suitable number of sets of 'cuts' for the engravings. The large printing cylinders will therefore perform two-thirds of a revolution while the smaller or type-cylinder is making one complete revolution.

"My next improvement relates to the inking apparatus, which is used in conjunction with the large printing cylinder, and consists in the use of an increased number of inking rollers and distributing rollers, so that the engravings may be more perfectly inked than heretofore. The ink, as is usual, is transferred by a vibrating roller from the ductor or fountain roller to the first distributing cylinder, from which it is taken by two rollers, and is deposited on a second distributing cylinder, to which an endway motion is given by means of any suitable mechanism. The ink is thereby evenly distributed over the surface of this second cylinder, from which it is transferred by other rollers to two other distributing cylinders, in contact with which four inking rollers rotate, and take therefrom the ink, which they transfer to the printing surface. In this manner the 'cuts' or engravings are plentifully and evenly

FIG. 89.—THE INGRAM MACHINE.

R 2

supplied with ink, and good impressions are obtained therefrom."

The machine itself is constructed upon the rotary principle, with which we are already familiar. The web of paper at one end of the machine is fed at once between the letterpress and impression cylinders. Before it is suffered to pass between the next pair, containing the electrotype plates of the engravings, it is calendered by two small rollers, and the impression of the type removed. It is now passed to the larger cylinders, upon one of which the electros are fixed in duplicate or triplicate, the slower rotation of these cylinders being compensated by their larger diameter, so that their surface speed is the same as that of the type cylinders. From thence the web, now fully printed on both sides, passes under a perforating knife, the sheets being finally separated by snatching rollers as in the Walter press. Lastly, a folding machine is attached in duplicate, in order that the folding may be done at a slower pace. The machine is capable of turning out 6,500 perfect sheets per hour, folded.

One marked advantage of this machine is found in the rigidity of the cylinders, and the consequent thinness of the overlay required, thus saving much time in making ready. The inking apparatus, as described in the specification, is almost perfect in its distribution, while the size of the outer-forme cylinder, on which alone, as a rule, illustrations are placed, does away with the difficulties caused by the excessive curving of the electrotypes, and consequent enlargement of the spaces between the lines in the engravings, the plates requiring very little curvature to fit the large cylinder used.

The original machine is now employed on the *Penny*

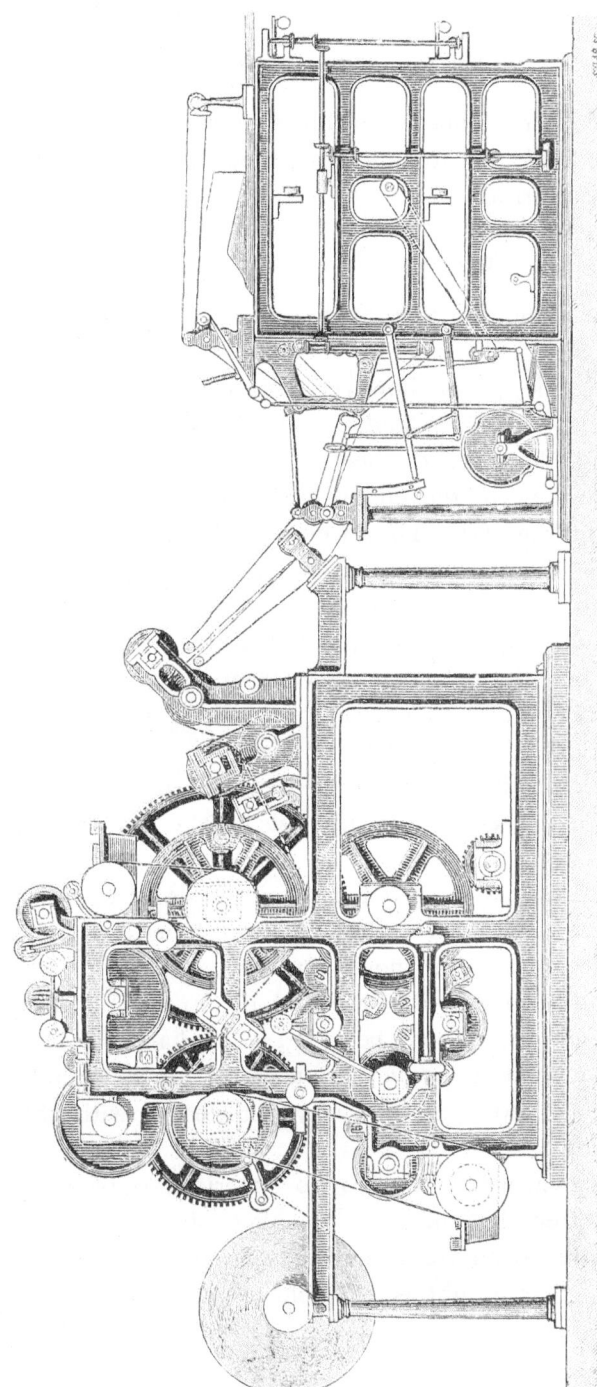

FIG. 90.—INGRAM MACHINE: SECTION.

Illustrated Paper. These machines are built by Messrs. Middleton, and the folding apparatus attached to them by Messrs. Harrild.

ANGLO-AMERICAN ROTARY.

This machine differs from the other rotaries inasmuch as it has neither grippers nor tapes to carry or guide the sheet. The paper, however, traverses its route from the reel with accuracy, and is cut to size by automatic means, and the printed sheets are laid one above the other on a table at the end of the machine, at the end farthest from the reel. The sheet is not separated until it is printed. Formes of various sizes can be printed on this machine by an easy adjustment of two cog-wheels. The supply of paper is so regulated as to give the exact length to be printed at each revolution of the cylinders by a simple yet ingenious motion, which it would be difficult to describe fully without sectional diagrams, but it is capable of giving out sufficient paper at one motion to print a sheet either the full size of the cylinders or any size smaller.

The impression is regulated on a simple plan, by the turning of a wheel, even while the machine is running. Thus, if a reel of paper should run dry, more impression can be readily given with ease and certainty ; or, if too damp, the impression can be eased immediately. These impression cylinders are very rigid, yet can be operated on at once without in any way interfering with their solid bearing.

The framework of the Anglo-American is of extraordinary strength, and the blanket and impression cylinders are so adjusted that they balance each other ; thus their great weight is not felt to any extent either by the substantial frame or by the engine that drives them.

FIG. 91.—THE ANGLO-AMERICAN ROTARY.

This machine is constructed to print a sheet 54 in. long by 41 in. wide, and its total length, including the delivery but exclusive of the reel, is 12½ ft., its width 7½ ft., and its height is under 7 ft. The printing machine proper is *but five and a half feet long*, which is only twelve inches longer than the largest sheet it prints ; and as it stands on a level floor, no pit is required. The two plate cylinders are geared directly with each other on one side of the machine, and on the other side they are geared each with its own impression cylinder, thereby insuring as accurate register as possible. These cogwheels are larger than the cylinders, so that any backlash may be minimised, it being necessary to guard against this when the impression is finished and the backs of the cylinders oppose each other, at which moment there is no impression being given.

Stereotype or electrotype plates must be used, which may be fastened in any position on the cylinders, there being no screw-holes nor grooves on their surface, side pressure only being required. The inking arrangements are perfect, always excepting that, as in all rotaries, the rollers go only one way. The ink has to pass between two milling rollers while still in the duct, and two beyond. Besides keeping the ink alive, this grinding helps to warm it, which is a matter of great importance where rapid printing has to be done during cold weather. The rollers are driven by gearing, thus preventing skidding. The sheet travels through the machine in a level direction, and the delivery, including the knife for cutting the sheet from the reel, is designed for a high rate of speed. The knife can be arranged to make more cuts than one to each revolution of the machine. If a folding machine is needed, it can be attached with no more alteration than wheeling away the flat deliverer and substituting the folder.

The Anglo-American machine can be made to print illustrations fairly well, and the delivery being wheeled away, the workman has easy access to his overlays.

ON STEREOTYPING FOR ROTARY MACHINES.

Before leaving the subject of rotary machines, it may be of general interest to give a brief description of the method employed by the stereotyper to produce the plates suitable for this class of machinery.

The pages are received in the foundry from the composing room, and locked up in strong wrought-iron chases. All the stereotyper has to do is to reproduce each page in one solid piece, or plate, and leave the type clean and uninjured. After a light rub over with an oily brush, layers of paper (thin tissue, and sugar, or blotting) are placed on the face of the type, and pasted each to the other, with a mixture of flour, whiting, and starch, or size, according to the whim of the stereotyper ; this, when thick enough, is beaten down into the spaces between the words and lines with a bristle brush, and the pasted paper, or " flong," as it is technically termed, is indented by every letter in the page. Sometimes a rolling-press is used instead of the brush, but the result is not so satisfactory. The forme, with its flong adhering to it, is next put under the platen of a screw press, the underside of which is heated either by the fire from the metal-pot furnace, or by gas-jets, and here the flong is pressed well down on the face of the forme. The flong having dried into a matrix (the pressure preventing wrinkling) it is lifted off, and presents an accurate mould of the type.

Another method of drying the flong into a matrix

is to peel it off the face of the type while it is damp, and dry it without the assistance of the forme, usually in a flat box of heated sand, which allows the flong to shrink while drying, so that, in the instance of advertisements, *more can be inserted* than could otherwise be got in, these advertisement pages being made up longer than the others. For an instance of this we may point to, among several daily papers, the *Daily Telegraph*, whose first page is, we believe, thus manipulated.

This matrix, having been brushed over with French chalk, is now ready for the casting-box, and is placed in position; being flexible, it readily lends itself to the curvature necessary to form a plate of the radius of the plate-cylinder. In the case of book-work and most publications, the plates, when cast, are kept flat, and finished so, but we are dealing here with plates for rotary machines, which must be rounded, to fit their respective cylinders.

The casting-box is semicircular, and, with that exception, like all other casting-boxes, so that when screwed up and tilted it leaves a space for the hot metal to run in, which gives the plate its exact substance. The metal being ready, and of the proper temperature, it is poured from a ladle into this gap, where it runs down, pressing the matrix close to the wall of the box. The matrix cannot burn, as it is prepared with whiting, which renders it almost fire-proof, and no air can get to it. The casting-box is then lowered, and opened by uncoupling the clamps and screw, the plate drawn out, and the paper mould peeled off. A curved *fac-simile* of the forme of type is thus obtained; the matrix, being little more than browned, is still fit to go into the casting-box again, if, by some accident, the plate is found to be damaged, or duplicates are required.

The plate is next placed on a cylinder of the same diameter as the cylinder on the printing machine, and the sides and ends trimmed, the piece of surplus metal, called in the stereo-foundry a "mouthpiece," being removed by means of a revolving cutter. The plate has usually, on its under-side, several short ribs of metal, cast in one piece with itself, and by laying this plate, face downwards, in a trough, and passing either a sharp-edge or a "rubber" over these, the plate is made of an even substance.

The latter process renders the plate tolerably even. Very little making ready or patching is done. In fact, in the majority of cases, the printing machine is started within a few minutes of the plate being adjusted, there being no time for improving the impression beyond the facilities afforded by the impression screws.

CHAPTER XIX.

ON BOOK-WORK MACHINES.

One-sided and Perfecting—The Class of Work they are adapted for—Speed—Various Merits described generally as far as Inking Arrangements are concerned.

BOOK-WORK machines are generally understood to be either the Wharfedale (in which class we include the Bremner, *Graphic*, the Reliance, and the double-platen machines) and the perfecting—the Anglo-French and the large cylinder Middleton, Dryden, &c.

While not disputing for one moment the distinct merits and capabilities of the platen, we would say that for the best class of cut book-work the Wharfedale will be found preferable. For many years the first-named machine was deemed to be in every way superior to the latter, but this was largely owing to the defects that existed in the construction of the Wharfedale, and also to exaggerated, old-fashioned prejudices of the machine-minder. It may be here stated that one of the chief difficulties with the original single-sided cylinder machine was the slurring But when the makers began to understand that this result was greatly owing to the defective cutting of the racks and geared wheels ; and the printer, that blankets were not absolutely necessary for cylinder covering, the slur soon disappeared. Again, more attention was ultimately paid to the inking arrangements, and now nothing that a platen is capable of can approach it for turning out superfine cut-work. It

may be urged that in the case of the Napier platen the
rollers pass four times over the forme between every
impression. This is perfectly true ; but it must also be
added that the speed at which the rollers travel over
the forme is very much greater than with the Wharfedale, and therefore the inking is not so effective, while
the rollers sometimes suffer in consequence of the quick
movement.

There is, perhaps, even at the present time, a time-
honoured prejudice with some, that the flat impression
similar to that of the hand-press, is necessarily superior
to the cylindrical. But although we in no way wish to
depreciate the capabilities of the platen machine, which
prior to the present make of the Wharfedale, was employed in almost every first-class London house for cut
and superfine book-work, we venture to assert that now
it has been conclusively demonstrated that the single
cylinder is capable, in the hands of a skilled man, of doing
superior work, besides being quicker and more economical in its working.

The platen, firstly, requires four boys—two at either
end, and although one machine-minder is supposed to
take charge of both ends, when it occurs that the printing of each forme is completed at or nearly the same
time, one set of boys are idle until the other end is
started, except, of course, another machine-minder is employed. The speed of this machine is, with both ends
combined, only equal to a Wharfedale, while the size
generally in use was no larger than double demy. A
Wharfedale capable of taking a forme of the same
dimensions is considered a very small machine — the
majority being at least quadruple crown—nearly double—
very many quadruple demy, while larger sizes are
often constructed. Thus we find that with half the

number of boys we can turn out double the quantity of work, as the platen, with both ends and a service of four boys, will print 1,000 sheets of say double demy per hour, while the Wharfedale with two boys, or even one boy, if the taking-off apparatus is employed, and no interleaving is required, will print the same number of a sheet twice the size. Of course platen machines are constructed to take a larger sheet than the one mentioned; but when the platen has a greater surface, difficulties in the working increase, especially with old machines, as the connection with the beam is immediately in the centre, and one side is apt to touch before the other. This, of course, does not apply to the Napier, as the platen head is held fast on either side by steel rods, as already described.

With regard to the driving power required, the old platen takes at least three times that of the Wharfedale. This will be readily understood when it is considered that in the case of the latter the impression is given simply at a point along the cylinder, which may be said to be half-an-inch in width, while the whole " nip " is administered at once on the platen. This proves to be a severe strain, both upon the engine and shafting ; and while counter-shafts are invariably adopted, or should be, so as to lessen the lug upon the main shafting, 4-inch driving-straps are used to communicate the necessary power. A workman may, with little effort, pull an impression by turning the fly-wheel of a Wharfedale by hand, but this is out of the question with the old platen machine.

In discussing the merits of these machines, the probable amount of repairs and consequent stoppages must be taken into account. The construction of the platen is not so complicated as that of the best class of

Wharfedale, but the cost of keeping the former in thorough repair is certainly greater. This is owing, not so much to the machine itself, as often to the carelessness of the boys employed. Unless the taker-off is well trained, a frequent "double-up" of the tympan and frisket is probable. If the layer-on is not careful as to the time when he strikes on, or to allow the pin to fall into the travelling slot, a breakage may result. To this we must add that the operation of both laying-on and taking-off is more difficult than on the cylinder machine.

From the foregoing it will be gathered that the Wharfedale machine is much to be preferred for quality, speed, and economy. Originally it costs less, it is less expensive to drive, the production is greater, and in the hands of men such as are at the command of all master printers who choose to offer sufficient remuneration, work can be produced equal to the finest sample of hand-work, with this advantage, that even colour may be relied upon—a quality not always identified with press work.

The machine capable of producing the next best quality of work is perhaps the Anglo-French, and, in fact, we have seen really first-class magazine illustrated work turned out by this machine. We must, of course take into consideration the class of workmen employed. As every overseer knows, one man will, on an Anglo-French, produce infinitely better results than another on a Wharfedale. But, given two first-class men, while the single-sided machine will undoubtedly be more successful than the Anglo-French, the latter will do really artistic work. But it must not be forgotten that the man with the perfecting machine has at least double the amount of work to perform. Both inner and outer formes have to be inspected continually, and thus trifles are apt to escape notice that do not in the case of the man who has simply one forme to attend to. This, together with the attention required in the supervision of a double set of boys, and the keeping in order of a double set of rollers, sometimes materially affects the general quality of the work. Again, the cylinders and formes of this machine are far more difficult to reach than in the case of the Wharfedale, and a slight defect which with little trouble or effort could be remedied on the latter machine, is sometimes allowed to go in the case of the Anglo-French; not that it should be allowed to pass, but it is frequently intentionally overlooked. When it is desired to attend to the outer forme, two taking-off boards have to be lifted, the work removed, &c., and this occasions a serious delay, which, if multiplied in the course of a day, would naturally interfere with the number printed; whereas in the Wharfedale the forme can be exposed in a moment, while the cylinder is un-encumbered with tapes, &c.

Perfecting machines are of course eminently adapted to long numbers. If the number, however, be, say, under 5,000, no matter what class of work, we are inclined to think that the Wharfedale is the most economical to employ. It is a well-known fact that it takes about one-third less time to make ready a single forme on a one-sided machine than on a perfecting, because the facilities are much greater; added to which the output is, or should be, at least twenty per cent. more. This fact is becoming generally recognised, as the demand for the Wharfedale in general printing offices is increasing every year.

Neither must we forget, in calculating the cost of these respective machines, that when set-off sheets are used on the Anglo-French four boys are required—two for the work and two for the set-off. Set-off paper with illustrated work is absolutely necessary in the case of the perfecting machine, while it may sometimes be dispensed with on the Wharfedale, because in the former case the inner forme is necessarily wet, being taken immediately on the outer-forme blanket. On the single cylinder the probability is that the first forme is comparatively dry before perfecting commences. If, however, there are heavy cuts, it is advisable to interleave, but this only takes one boy.

For speed, the ordinary perfecting machines of Messrs. Middleton, Dryden, or Dawson are to be preferred. Several attempts have been made at various times to increase the output of the Anglo-French, but they have not been attended with much success. As before described, the upright spindle has, to this end, been substituted for the horizontal rack, but immediately the machine is run, say, at the rate of more than 1,000 copies per hour, the continuous supply of fresh set-off paper, one of the advantages of this machine, becomes impossible, as the boy is unable to fly the latter from the surface of the printed sheet before another is thrown upon the taking-off board. Continuous set-off sheets, although sometimes used, are not at all to be recommended, as they are apt to become torn by the grippers of the inner cylinder.

With the large cylinder machines this objection does not hold good, as speed but little affects the run of the continuous bands. These machines are generally constructed to print between 1,400 and 1,750 per hour, and although the makers profess that they may safely be driven at a higher rate of speed the laying-on and taking-off become a source of great difficulty, especially when the paper is thin and damp, as it has an inclination to double up on the taking-off board, necessitating a stoppage to straighten the heap.

S

As far as the quality of work is concerned, this machine may lay claim, in the hands of competent men, to being able to turn out good ordinary magazine work —not the high-class periodicals, with "close" cuts, now becoming general, but what we may be excused for calling second-class illustrated work. It is by no means desirable to select work with solid cuts—first, because the rolling is not so perfect as on the machines previously mentioned ; furthermore, the speed materially interferes with a perfect deposit of ink ; and the set-off sheets become so speedily soiled that the stoppages necessary to renew them reduce the output.

The inking on this machine is certainly not so good as on the others before mentioned. This is because the tables have not so long a travel, the consequence being that only three rollers entirely cover the forme, the fourth, nearest the cylinder, "turning" on the entry.

For ordinary work, however, as before stated, they are especially fitted. Their construction is so excellent and simple that they require very little expenditure in the matter of repairs. As described in the chapter devoted to this class of machine, the bar is generally preferred for speed, while the gripper is more reliable for accurate lay.

Part III.

———•◦•———

CHAPTER XX.

PRACTICAL PRINTING—MAKING READY.

Preparation of Machine—Blocks and Catches—Principles of Making Ready—Type and Stereoplate—Underlaying—Making Register—Overlaying—Inking-up and Starting—Making Ready on Single Cylinder, Tumbler, Two-colour, Platen, Napier Platen, Ordinary Perfecting, and Anglo-French Machines—General Remarks.

BEFORE laying a forme on the machine, it is absolutely necessary that the machine itself should be in a fit and proper state to receive it. No amount of making ready can insure success if every detail has not been properly attended to, in the shape of oiling cleaning up, &c. Presuming a workman to be laying on the initial forme of a job, and that he is of course cognisant of all the details of his machine, he should satisfy himself that every bearing is properly lubricated, that all the tapes are in running order, that the ductor is free from grit, and that the coffins are carefully scraped and cleaned, so that nothing may interfere with the impression when the forme is laid on.

Before entering into details, we may state generally that the object of "making ready," or preparing a forme in such a manner as to insure satisfactory results, consists, first, in so placing the forme as to allow an even margin, and underlaying, or so adjusting the inequalities of the plates from underneath as to enable the rollers to touch evenly the entire surface. The *impression* could really be equalised by means of overlaying, *i.e.*, by placing

S 2

pieces of paper upon the cylinder in such a manner as
would force the sheets to "dip" into those places that
were lower than others, and by cutting away at the parts
where the impression is excessive. But if this system
were adopted, it would be found that although the impression might appear to be absolutely level, many parts
would show up of lighter colour than others, this arising
not from the impression being unequally adjusted, but
from the fact of the surface of the rollers not having
touched those parts with sufficient pressure to deposit
the necessary quantity of ink.

These latter remarks do not, of course, apply to
type formes, or movable matter, as these should never
be required to be underlaid. Type pages are invariably
of the same dead level, and nothing more than judicious
planing should be necessary in the preliminary process.
Sometimes, however, it is found that the coffin or table
of a machine has sunk in the centre, rising to either side.
In this case a few sheets of paper may be used, de-
creasing the number towards the edge. But on machines
of modern manufacture this is seldom necessary.

When the first forme of a work is sent down to
machine, the plates are generally imposed, or mounted
and fixed upon a solid bed. These beds are made
of different materials—iron, mahogany, or lead stereo-
blocks, the latter being the most favoured. We very
much prefer the small blocks which consist of leaden
cubes, flat on the top and hollow underneath. To
simplify the making-up of a number into a page of any
given size, they are generally cast in two sizes, the large
one being $2\frac{7}{8}$in. by $1\frac{1}{2}$in., and the small one $1\frac{1}{2}$in. by $1\frac{3}{8}$in.
Should the most convenient combination of these be too
small for the plate, pieces of solid metal furniture of a
nonpareil, pica, or double pica are dropped between

the central blocks, and sufficient leads are used to exactly adjust the entire surface to the required size. The plate is secured to the whole by means of brass catches, two or more being placed at each of the four sides. Every block has a small space in the centre of three of the sides, with a small hole to admit of the pin of the catch fitting exactly. The back of the catch when in position falls exactly flush with the side of the block. This can be readily understood from the following diagram. It will be noted that the top portion of the

Fig. 92.—Small Stereo Block and Catch.

catch is bent inwards. The lip falls over the bevelled edge of the plate, and firmly secures it to the blocks.

An improved catch, the invention of M. Corsain, is to be highly commended, as it admits of the plate being removed from the block without unlocking the forme. It consists of a clump or small piece of furniture, two picas in width, with two grooves, one being dovetailed to admit the catch, and the back one holding a slight spring, which presses firmly to this catch, pressing it to the bevel of the plate. When it is desired to remove the catch, a key is pressed down between the spring and the back of the catch, when a small pin fits into a slot of the former, and may be easily withdrawn.

It is simple in construction, as will be seen from Fig. 93. Messrs. Powell of Ludgate Hill, are the agents.

Sometimes special metal blocks are made, much larger than those mentioned, which are preferable when there is a continuous supply of work of a definite size. They are cast to a given length, so as to exactly suit the page, allowing, however, sufficient space at the

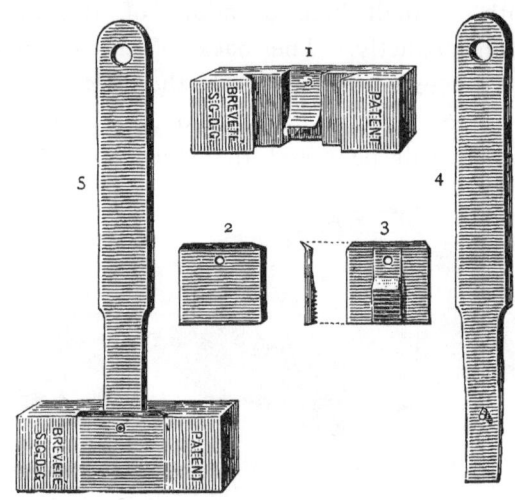

Fig. 93.—Corsain's Improved Catch.

1. Clump with spring; 2. Catch (front); 3. Back and side; 4. Key; 5. Key in position for withdrawing catch.

extremities for a small metal clump, with aperture for the catch.

We do not propose to enter into the matter of imposition—the laying down of the plates and adjusting of the spaces between the pages to allow of proper backs and gutters. This more properly belongs to the compositor's branch of the business than to the machine-minder's, although it is frequently found necessary to alter slightly these spaces by means of leads and reglet to insure perfect register, as we shall afterwards explain.

We will first deal with the making ready on a

Wharfedale machine, the remarks applying of course generally to the Bremner, Dawson, Payne, and similarly constructed cylinder machines.

Prior to laying-on the forme, the cylinder should be properly prepared, the blanket stretched, and everything in fit and proper condition to pull an impression. Primarily it is necessary to put round the cylinder a foundation, as it were, upon which the sheets, &c., can be fastened. Calico is nearly always used for this purpose. A piece must be obtained nearly as wide as the machine cylinder, and sufficiently long to extend entirely round its surface. One end is fastened to a fixed bar with a ratchet motion at one end, and the other to a loose rod which may be lodged behind a series of pins beneath the gripper-bar. By means of a key the calico can be stretched tightly round the cylinder, and held firm by the ratchet wheel. Upon this foundation are placed the cylinder sheets. These should be slightly wider than the forme to be printed. Hard paper may be selected, and, if possible, it should be rolled, as there is then less difficulty in obtaining a perfectly smooth and unyielding surface. If the forme be (say) quadruple crown, paper of weight (say) of 60 lbs. to 70 lbs. will be found eminently suitable. Take (say) eight sheets,* lay them on the board flat, and "fan" them out so that the front edge of each lies behind the other by about one inch. Paste the whole of these laps, and carefully pull each sheet on top of the preceding, thus securing them all firmly at one edge. The greatest care is necessary that every sheet be perfectly smooth and flat on top of the underneath one, as if the pasted edge is at all cockly, a crease

* The number of sheets must necessarily be regulated by the thickness of the blanket employed.

will result. When all are firmly fastened together, allow the boy to turn the fly-wheel of the machine, stopping the cylinder when the gripper aperture is uppermost. Force open the grippers, and after pasting the under-side of the already pasted edge of the series of sheets, press the entire length firmly to the calico immediately inside the cylinder opening, and so make the whole fast. The cylinder may now be allowed to complete its revolution, and the loose end of the paper be smoothed round its surface. We have now a solid and firm foundation, upon which may be affixed the overlay, &c. Where the hard-packing system is adopted, the cylinder may be considered to be almost finished ; but we shall first deal with the making ready with a blanket.

This material is really the finest cloth, varying, however, in thickness. For fine work the thinner kind is used, but for newspaper and common work that of thicker texture is adopted. The reason for using the thicker material is that it will more readily adapt itself to the inequalities of the surface of the forme. But the overlays are by no means so effective, as the material intervening between the patched sheet and forme prevents its "telling ;" whereas when the overlay is allowed to come closer in contact with the surface of the forme the thinnest piece of paper affects the impression. The blanket, like the cylinder sheets, should be wider than the forme. A series of small holes must be made in one end, through which the blanket-rod can be put. This rod is lodged behind pins beneath the grippers, and then pulled tight round the cylinder, being pinned at the other extremity.

When the forme (or formes) of plates is received by the machine-minder, it is, after finally wiping out the coffin, laid on to the table, care of course being taken that the proper end is placed to the gripper edge.

A gauge is generally kept by which the correct
" pitch " can be ascertained. This is calculated from the
edge of the ink table. Of course, be the size of the
sheet crown or quadruple crown, it does not in the
least affect the position of the " entry," as the grippers
take the sheet at a fixed point, so that the forme
must be laid, so far as the part to be first printed is
concerned, exactly in the same relative position, the
waste space, if any, being at the other end. Care must
be taken, however, that the forme be placed in the
centre *across* the machine. In securing the chase, every
precaution should be used that it is so arranged as not
to " spring." The furniture must be solid, and, in the
case of small formes, galley chases may be used to fill
up. The locking-up should be done on the leaving end
and on the near-side, the other sides being filled up
squarely and solidly with furniture.

First, it is necessary to ascertain if the cylinder be
properly adjusted on its bearers. Of these there are
two sets running inside the table racks. The outside
(the widest) are for the cylinders, while those running
flush and parallel are those upon which the rollers
travel. As will be readily understood, the bearers give
solidity and support to the cylinder—either side resting
and running upon them. Therefore it is necessary to
so adjust the cylinder by means of the impression
screws, on either side, as to *rest* but not unduly press
upon these hornbeam slips. The amount of pressure
may be ascertained by laying a slip of paper on the
bearers and running the machine, when it can be easily
decided if it is necessary to raise or lower the cylinders.

We will first deal with a sheet of plates—stereo or
electro, as the case may be—containing no cuts.

If the forme be imposed in folding chases, see that

the two thin bars are placed together in the centre of the coffin, and perfectly flush.

When the forme has been properly centred, a rough impression must be pulled, and we shall at once ascertain what is necessary to be next done. It will generally be found that the height of a plate, mounted upon stereo blocks, is slightly lower than type. Probably the impression will be found to be extremely faint, and if the blocks and plates be in good condition the general condition of the forme will be the same. Supposing the cylinder to be well secured upon its bearers, three alternatives are open—a piece of glazeboard may be placed under the entire forme, extra sheets may be put round the cylinder, or the cylinder itself can be lowered. It is not desirable that the cylinder should be forced down, as this is apt to strain the machine. The first expedient is the best, quickest, and most effective. In dealing with the cylinder covering, it should be always borne in mind that practically *every sheet placed round the surface increases the circumference.* If this is done indiscriminately, it will be seen that the impression surface is made greater than the face of the formes—resulting in a slur—one surface moving at a different rate of speed from the other. Of course a single sheet of paper is a small item, but the multiplication of them leads to this result. In this case a portion of the dressing must be removed, and the equivalent in thickness placed underneath the plates.

Electros will naturally be preferred by the printer to stereos, not so much because the surface is harder and sharper, but because, in the majority of instances, more care is taken in the finishing. If we take a sheet of plates we shall certainly find that the electros yield a very much more satisfactory impression than the stereos. This is owing greatly to the fact that the stereo plates shrink un-evenly in the cooling process. And although they may be planed on the *back* in a similar manner to the electro, the face is slightly uneven, which is almost im-possible in the case of the electro. The making ready of a sheet of plates is a very different matter from that of "movable," or type, for, as before explained, no underlaying is necessary in the latter case, while with plates the success of the printer's efforts entirely depends upon the state in which he leaves his forme prior to overlaying. It will be found when the initial impres-sion has been pulled that some pages are lower than others. Whether this is owing to the defective planing of the plates, or the worn condition of the blocks upon which they are mounted, matters little. They must be *levelled up.* If the whole page is low, put a piece of brown paper or glazeboard under the blocks before commencing to patch the underlay. If only a portion of the plate is low, an extra piece must be put under the depressed part.

First, unlock the forme and plane it. A small piece of wood, say three or four inches square, having leather fastened on the under-side to prevent possible injury to the face of the plate, is generally used. The object of unlocking is to allow the blocks to fall squarely upon, and adapt themselves to the coffin—thus obviating all "spring." Compositors are prone sometimes to lock up very tightly, the result being that the pages are inclined to rise in the centre, yielding a false impres-sion. After planing, tighten the quoins with the fingers, or by gently tapping with a shooting-stick. Having obtained an impression of the forme, making sure that the whole generally is tolerably level, roll it with a hand roller.

When the forme is well rolled, lay a full-sized sheet over the surface and allow the machine to run, striking-on the cylinder. It is desirable that the machine be fully struck-on, as otherwise the impression may be inaccurate. Lift the sheet from the forme and put it upon a flat board, having an elevation of say 45°, of course in a well-lighted position. The inequalities of the impression will soon be perceptible. Our object now is to so adjust the underlay as to raise the depressed portions of the plate and lower the high parts, that a general level resistance may be exerted. It will be understood that it would be of very little service to use very thin paper for this purpose, as we have a solid piece of metal three-eighths of an inch thick to deal with. Paper, say, equivalent to double-crown 50 lbs. is about the substance adapted to this purpose. Practice only can determine the *amount* to be placed on the various depressions—whether two, or even three pieces, one on the top of the other, are necessary to effect the desired change. Frequently it will be desirable that several pieces be affixed over the largest portion of the plate, reserving the final touches for the second overlay. We have to *raise* the low parts, the high portions being simply cut out by means of a sharp scalpel. After having patched the sheet, the pages must be cut up separately, trimming each flush to the edge, so as to prevent the paper protruding beyond the bevel of the plate when placed underneath. It will save time to ascertain exactly the position of the pages, compared with the forme, cutting them apart in rows, and as each successive set is done, placing them on their respective plates face downwards. When this is finished, unlock the forme, raise the plates from under the catches, and paste the underlay in position underneath. The paste should be used sparingly, and should be entirely free from lumps. Sometimes the underlay is simply laid on the blocks, and the plate placed on top. This is rather an unsafe method of proceeding, as it may be slightly shifted when replacing the plate under the catches. The best method is to paste a small portion both top and bottom, taking the plate in the hand and adjusting the headline first. It must be borne in mind that when the forme is worked off, the underlays have to be entirely removed, so that it saves a deal of trouble ultimately when only just sufficient paste is used to insure the underlay adhering securely. It may perhaps frequently be noticed, when the plates are required for reprinting, that lumps of paper are firmly fixed to the under-side. This is owing to the over-free use of paste, or to the carelessness of the machine-minder, when lifting, in not removing the entire underlay in the original working. It should be understood that although a set of plates may have been most carefully and success-fully levelled up in one case, if the same underlays were tried on another machine, with or without the same set of blocks, the impression would be woefully bad, owing to altered circumstances, as far as bed and cylinder of machine were concerned.

Having affixed the underlays to their respective pages, and properly placed the plates under the catches, again tighten the quoins, and plane down. Everything being clear, place a sheet of paper over the face of the forme, and allow the machine to run several times, striking the cylinder on. This should be done slowly, in order that the plates may be forced down upon the underlays, and assist them to adapt themselves to the altered condition of pressure. In some cases one underlay is enough, the pressure having been so well adjusted as to render a second unnecessary. We will in the present case suppose that the pressure is not sufficiently even to allow the overlaying to be proceeded with. When the machine has been allowed to run for a few minutes, during which time the minder should be careful to see that the waste sheet entirely covers the forme, to prevent blacking the blanket, again roll with hand roller, place a clean sheet upon the surface, and take an impression. When pulling the sheet for final underlay, a duplicate must be printed for the purpose of revision. This, of course, is sent to the reader for testing imposition, &c., as although the forme may have been sent down to machine perfectly correct, pages may have become transposed in underlaying. It will now be seen what effect the patching has had, and although the impression may not be absolutely level, every portion of the page should be clearly visible at the back when held up to the light. It is not desirable that much should be placed on the second underlay—indeed, it should not be necessary. If it is, the first patching has been badly done, or the plates must have been in a very bad state.

As before, patch the light places and cut out those portions still too high. Paste on the top of the first underlay, and again secure the plates. Inasmuch as it is presumed that the underlaying is now completed, the forme should be firmly locked up and well planed.

While on this subject, we must refer to the practice of sometimes placing too much underlay beneath cuts. It is a common expedient to paste two or three pieces immediately beneath a solid or close portion of the cut, in order to insure that part receiving additional pressure. This is absolutely unnecessary, for, as has been already mentioned, the object of the underlay

is to *level* the plate, that the rollers may touch every
part. If an additional piece of paper is placed under a
portion of a cut, regardless as to whether or not it is
absolutely necessary for levelling up, it interferes considerably with the subsequent process of overlaying; in
addition to which, it is impossible to bring up a defined portion without slightly affecting the surrounding
edges; and these latter, therefore, are frequently injured
by so doing, and no amount of subsequent cutting
away of the overlay will remedy the defect. We
would impress this upon the young machine-minder,
as in good cut-work we have known a great amount
of trouble to be caused by the injudicious underlaying
of solids, done upon the supposition that the overlaying
would be rendered easier.

When the forme is sent to the machine room, it is
generally carefully gauged; but it will be found that
several plates will have to be slightly moved and readjusted, in order that their impressions may exactly
back each other when registering. Straight-edge the
pages, and endeavour, as far as possible, to get them
into position before proceeding further, or the work
entailed in the subsequent alteration will be greatly
increased. Fix the spurs or points in the back of the
gutters, one in about three inches and the other two
inches. By thus varying the distances from the edge,
the laying-on boy may easily detect if a sheet be turned
the wrong way round when perfecting. These points
may be secured by melted compo or tacks, but care
must be exercised that they are firm, otherwise the
vibration of the machine, together with the action of
the rollers, will loosen them in the working, and they
will pull off, and possibly cause a batter.

If the work is half-sheet work, *i.e.* perfecting on

itself—each copy when backed forming a duplicate of the entire forme—one half must be registered to the other, as it is not desirable to shift the pages on either side. The portion to be registered to should be gauged very truly before finally locking up.

Under the laying-on board is a movable frame with slides, into which points can be screwed, one on either side, to correspond exactly with the spurs on the chase. In the first place pull a sheet, and lay it on the board face downwards, taking care that the entry is perfectly correct, *i.e.*, not reversed. The adjustable points can be now so fastened as to exactly fit the holes already pierced in the sheet. Turn the sheet face upwards, and place it upon the spurs. Allow the machine to run, taking an impression. By holding the sheet up to the light the success of the imposition will be tested. It may probably be found that the sheet is entirely out on one side. This may be remedied by slightly tapping one point on the side, which will move the *whole* of the paper in the desired direction. Pull another impression, and by carefully manipulating the points we shall, if proper care has been exercised in the gauging, bring the plates generally into register. Several pages, however, will possibly require to be slightly altered to secure a dead fit. Unlock the forme on the one side, and, by holding the sheet to the light as before, it can be nicely calculated if a thin or thick lead is necessary to be inserted at any point to exactly square the page. It will be remembered that we are now dealing with half-sheet work. If the imposition be arranged as inner and outer, *i.e.*, the second forme to perfect on the first, the register would of course be made on the former.

Having received the revise from the reader, and supposing everything is perfectly correct, *i.e.*, that every page is in sequence, and that there are no batters, &c., we are now in a position to prepare for the overlaying. The whole of the foregoing relates, of course, simply to plates. If we were dealing with a forme of type, as before explained, no underlaying would be required; and after laying the forme exactly in position we should immediately proceed to overlay.

The inking apparatus, consisting of ductor, vibrator, wavers, and inkers, must now receive attention. Hitherto a hand-roller only has been used, but now we can " run-up " colour.

THE DUCTOR.

Some degree of care is necessary in the management of the supply of ink to the vibrator, otherwise the inking will be faulty; and we suspect that the composition rollers are often blamed as being unfit when the real cause lies in the unskilful way in which the knife, &c., of the ductor has been treated. The flow of ink from the ductor can be regulated to a nicety by the set-screws at the back, as figured on page 291, but in any case the "tommy" should be judiciously, not to say sparingly, used, especially in dealing with the drawback screws, otherwise the knife is apt to spring. It is far better to allow a small quantity of ink to be taken, and increase the speed of the duct roller, than to let the vibrator be supplied with thick ridges of ink, as perfect distribution in this case is almost impossible. Before putting the ink in the ductor, place the leaden "stops" immediately in a line with the gutters of the forme, for if ink be allowed to run on the table in places where it is not required, it "gathers" on the rollers, and is deposited in ridges along the head- and foot-lines, filling up the beards of the letters, and producing a muddy effect. These

T

stops may be made by pouring molten lead into the ductor between pieces of clay previously adjusted to the required width. When cool, any rough edges can be cut off. Every machine should be supplied with a set; also the vibrator can be cut to minimise the supply of colour at various points when necessary. First trim the ends of this roller, paring down the sharp edge, and lay it along the ink table under the duct roller, and immediately in the same position that it will eventually occupy. Carefully roll it along the table on to the first row of plates, and with a sharp knife cut lines to indicate the position of the gutters (where little or no ink is required). Take the roller off the machine, lay it upon the table, and cut away, between the lines marked, to the depth of say half an inch, removing the composition round the roller to the depth mentioned. Be careful not to cut too deeply, for if the incision is made too far in, we reach the stock, and by so doing weaken the adhesion of the whole of the composition to the barrel. If the knife of the ductor be not in good repair, the ink will probably flow too freely at various points, and it may be necessary to cut thin ridges along the remaining composition, thus still more reducing the surface presented to the ductor roller.

Although there certainly exists a very strong prejudice in favour of cutting the vibrator as described above, we do not think such cutting is absolutely necessary. If the knife of the ductor be perfectly true, the same be nicely adjusted, and the stops judiciously placed, then, given the determination to succeed, there should be no real difficulty in properly limiting the supply of ink to the vibrator. We know of machine rooms where the vibrating roller is rarely, if ever, cut—although the work is of a very varied description. This is certainly an

important item in the general economy, as, if the
vibrators are habitually cut, renewals of necessity must
be frequent ; whereas, when the vibrator is worked intact,
patent composition may be used—lasting for months.

After setting the ductor, both the cylinder bearers and

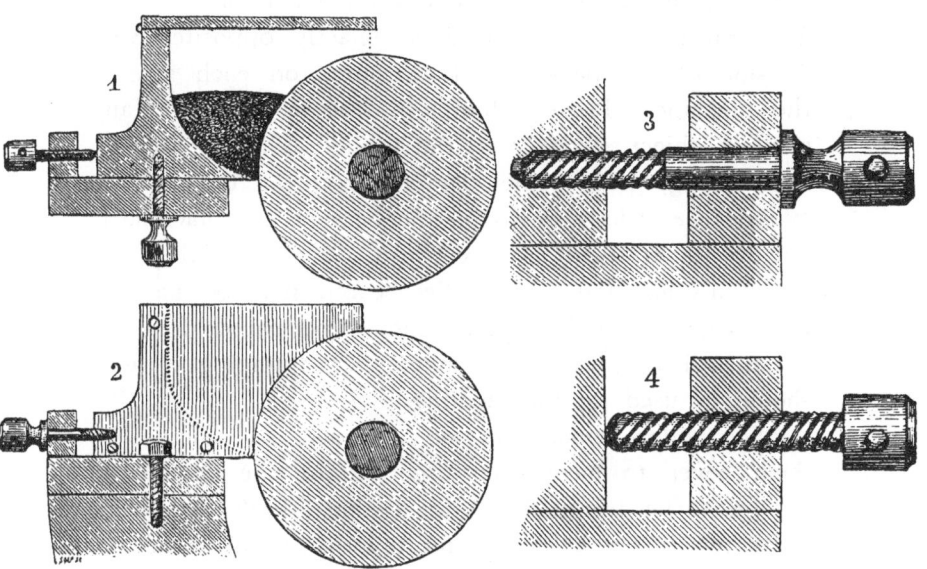

Fig. 94.—The Ductor.

1 Section of ductor, showing ink and position of knife ; 2 Ductor—end
view ; 3 Draw-back screw ; 4 Tightening screw.

the roller bearers must be "packed." The object of this
is in the first place to bear off the impression from the
edges of the pages, as the cylinder has an inclination to
dip in the gutters ; and to slightly raise the rollers at the
same point to prevent them wiping or unduly rubbing
which would tend to make them jump slightly immediately
they touched the sides of the pages—leaving a ridge of
ink, and possibly causing a "friar" immediately after. For

T 2

the cylinder bearers, wrapper paper is generally adopted. Lay a piece of furniture along the gutters of the forme, allowing the end to extend to the bearer, and cut a piece of the material mentioned about a pica each side wider than the gutter. Centre this carefully, and fasten to the bearer by melted composition or tin tacks. This must of course be done exactly opposite every division of the pages, on the bearers on each side of the machine. Prevent lumps in the fastening. Many printers object strongly to the use of cylinder packing, maintaining, with some degree of truth, that if the cylinder be strictly down on the bearers, the machine is apt to be strained by their use. However, they are commonly adopted. The roller bearers may be made of thin strips of leather, which should be pared down at either end to prevent the inkers from jumping. Tacks should be used for the securing.

The rollers must now be attended to. Supposing them never to have been used before, the edges will require to be trimmed. With a sharp knife cut away about two inches at either end, tapering towards the extremities. If it is found that the composition is in any way loose on the stock, hold the end in a gas-flame for a moment, and press on to the barrel. If this is not attended to, the material is apt to loosen, and the composition will leave the stock and drag on the table.

Wipe down both inkers and wavers with water; slightly warm if the weather be cold, to give them the necessary "tack." If patent composition is used, turps must be substituted for water.

Colour should be next run-up. First, so adjust the machine as to allow the vibrator to press flush to the ductor roller, and turn the latter round several times to insure the former being well covered with ink. Of course where there is an inking cylinder two vibrators are necessary, but it is only necessary to cut one.

Place the four wavers in position at different angles across the table. This will insure each pair travelling at different angles across the slab, facilitating the distribution of the ink. After seeing that everything is clear, and that nothing in the shape of loose pieces of furniture, &c., is lying on the coffin or surface of the forme, allow the machine to run for a few minutes, taking care, of course, that the cylinder is struck off. When a sufficient quantity of ink has been deposited and distributed on the table, stop the machine and place the inkers in position. Run the machine a few times, and feed-in a quire of waste, after which pull an impression on its own paper. It is advisable to take a few sheets out of the centre of the ream, especially if it be damped, that we may obtain a sound idea of the result of the underlaying. It will probably be found that the impression is heavier than anticipated, as dry sheets have previously been used ; whereas, paper when damped is softer, and offers less actual resistance. It may be necessary before proceeding further to slightly raise the cylinder, loosening the top set-screws and tightening the under ones. This, however, must be done very carefully, as it may affect the result of the packing on the bearers—necessitating sometimes another piece of wrapper being fastened on top of the original one.

We may state, with reference to the above, that it should be the endeavour on the part of the machine-minder to make it a rule not to interfere with the cylinder screws after the packing, &c., is fixed, unless absolutely necessary, as such disturbance is apt to cause slurs.

Having ascertained that the impression is generally satisfactory as far as the underlaying is concerned, the sheet should be examined on the face, that the inking may be scrutinised. If one portion is lighter than another, the screws at the back of the ductor possibly require readjusting. Turn the ductor roller several times, and pass the finger along, and if the ink is flowing unevenly it will most probably be felt. Supposing that it is desired to lessen the supply at a given point, the small or drawback screws should be loosened, and the long or set-screws be tightened. The former possess a shoulder, and the extremity being fixed to the knife, forces the latter away from the ductor roller when tightened. The former presses against the knife, and by so doing decreases the amount of ink escaping at the edge. As before mentioned, when once set properly it is extremely undesirable to interfere with the screws, as unskilful and too-frequent alteration of the ductor knife leads into all sorts of difficulties, especially in the case of a beginner. It must also be borne in mind that it is not desirable to tighten one screw without correspondingly loosening the one next to it, or the knife will, by unequal pressure, become bound.

It will sometimes be found, by passing the finger along the ductor roller, that thin ridges of ink appear here and there. This cannot be prevented by alterations of screws, being caused by an indentation or slight break in the edge of the knife. And inasmuch as it is rarely considered necessary to have the edge re-ground in consequence, a small piece may be cut out of the vibrator at this point, if the inking of the forme is affected.

In scanning the surface of the sheet it may be noticed that the edges of some of the pages are much lighter in colour than the rest. By turning the sheet and looking at the impression it can be easily seen if this is due to the want of pressure or to defective inking. Generally speaking, it may be attributed to the latter cause. The impression being fairly good, the defect is owing to the inkers not properly touching the forme at these points. The roller packing is too high, and must be pared down slightly at each end with the knife to allow the roller to touch the pages with necessary pressure.

OVERLAYING.

Having assured ourselves that the register is absolutely perfect, an impression may be pulled on the cylinder sheets, having, of course, reserved a good sheet for overlaying. It must be remembered that the position of the pages cannot be altered after the cylinder is pulled, as otherwise the impression on the cylinder would be deceptive. Therefore all the necessary alterations in the position of plates must be made prior to the impression being printed on the cylinder sheet. Allow the machine to run, that the forme may be properly inked, lift the inkers, take off the blanket, making marks round the sides so that it may be afterwards replaced *exactly* in the same position, and then allow the cylinder to make one revolution. By this means the exact position of the various pages is indicated on the cylinder. In overlaying, the sheet should be placed on a board at an angle, as before ; but inasmuch as there is a vast difference between the thickness of a thick and stubborn metal plate and that of a fine blanket, much greater care must be exercised in the patching. Thin paper, say 15 lb. demy, cut in long narrow slips, is well adapted, taking care that the paste is thin in substance, and also sparingly used. The heaviest parts will most probably be the edges, owing to the inclination of the blanket to dip into the gutters. These should be cut away, and the depressed portions equalised by the pasting on of small pieces of paper. The headlines and folios will also most likely be heavy, in which case they must be cut off. When the patching is completed, cut the sheet up into separate pages, and paste the overlays on to their respective places on the cylinder. This can be done with a little care ; but seeing that the overlay will be worse than useless if it is placed but a little "off," it is necessary that every portion be placed absolutely " dead " in position.

After having fixed the entire set of overlays, the blanket should be replaced, every precaution being taken that it be stretched exactly over the cylinder as before, the marks made prior to lifting being the guide. It will be seen how necessary this is when it is explained that the blanket, unless perfectly new, wears unevenly —some parts being thicker than others—and of course this would materially interfere with the evenness of the impression. Pin the leaving edge firmly, that it may not shift when the cylinder is in motion.

The forme having again been brushed out, and the inkers wiped down and placed in position, run a quire of waste, followed by one of the sheets which is to be ultimately used. If ordinary care has been exercised, very little should remain to be done. Reversing the sheet and holding it to the light, several inequalities may still be apparent. Consequently, another overlay may be necessary. Patch as before, thin set-off paper in this case possibly answering the purpose. While this is being done, the boys should be set to thoroughly " wash up "—wavers, inkers, and ink table being well cleaned ready for the final start. A letterpress forme should *never* require more than two overlays, and even when cuts are present a third should be unnecessary, supposing that due care has been taken at each previous stage. The patching of the second overlay should occupy a very short time, and after each page has been put up as before, there should be no reason why a start should not be immediately made.

It must be remembered that we have in the foregoing supposed that the forme is without cuts, and that a blanket has been used.

The paper having been placed in position, and the side mark adjusted, run a quire of waste and pull a sheet. The work should now be ready to start, but if any defect still exists it will be mainly owing to inattention to details. In the case of old plates or worn type, it will sometimes be found that, notwithstanding every care, the face of the impression is decidedly "scabby," which is attributable neither to defective inking nor impression. In this case a sheet must be "faced," *i.e.,* patched with set-off paper on the front where light parts appear, or cut out where heavy. It is not always necessary to put up an extra sheet on the cylinder for this purpose, as it may often be done by raising the blanket and putting pieces on, or cutting portions from the overlays already pasted up. Sometimes pieces are fastened on the outside of the blanket, but this is a practice to be deprecated, being both clumsy and slovenly. The blanket should always be clean, and ugly pieces of paper pasted over it are by no means a testimonial to the workman's skill in making ready.

A word or two with reference to the printing proper. When a machine is employed to do work, there is always a probability of the unforeseen happening. Success is entirely dependent upon each and every portion acting in a proper manner; and inasmuch as a printing

machine is somewhat complicated in construction, there
is always the possibility of something going wrong
This especially refers to the inking.   In nine cases
out of ten, if anything is to be found fault with in
simple letterpress formes, it is the inequality of colour.
Careful as a man may be in setting his duct, the work
will require constant attention.   Work will become too
full, or too spare, even after a considerable quantity
has been printed.   This may be occasioned by the ink
having become thinner, and flowing more freely, or the
composition of the rollers becoming either harder or
softer, owing to atmospheric conditions ; also by the
paper becoming drier than when first started.   The only
precaution against any mishap in this direction is constant vigilance on the part of the machine-minder.   It
is a good plan to have a colour-sheet constantly to
hand, which may be compared with the work as it is
running.   If the rollers are suspected, have them taken
out and sponged ; but if for some time inking has been
satisfactory, do not commence (as we have had occasion
to mention before) to meddle with the ductor screws to
any extent.   If this is done, there is little probability of
turning out satisfactory work.   If any radical alteration
is made in this direction, the probability is that the
workman will not be able to leave the ductor during the
entire working of the forme, and the ultimate result will
be anything but satisfactory.

Thus far we have dealt simply with an ordinary
forme of plates worked with a blanket.   Next to a
movable forme without cuts, this may be considered
the easiest to prepare.   Frequently formes are made up
of both movable and electro, in which case it will be
found generally that the blocks are slightly lower than
the type.   The former have of course to be underlaid to

the level of the latter, and the overlaying may require more care than an ordinary sheet of plates. With independent blocks, generally wood, there is sometimes an inclination to "bow," in consequence of the material having shrunk or warped from damp. In this case, the best plan is to lift the block and have it planed on the under-side, and pack level with a piece of glazeboard or wrapper.

As we propose to deal with the cutting of overlays in a separate chapter, we shall content ourselves here by simply saying that the whites and high lights should be cut out in the underlay, which will tend subsequently to simplify the work.

When "backing" or perfecting the forme, a thin sheet of paper should be fastened round the cylinder, to prevent the blanket from becoming soiled by contact with the side already printed. In half-sheet work, when the perfecting immediately follows the white paper, of course the ink is more or less moist, and soon soils the set-off sheet. When the latter becomes dirty it should be removed, and another substituted, or the side of the sheet just printed will speedily become marked.

In the case of sheet work, *i.e.*, when the inner and outer formes are worked separately, every precaution must be taken that the first forme be most carefully gauged and straight-edged, for if any of the pages are out of the square, the one that falls on the back in the second forme will have to be proportionally skewed, or the register will be defective.

Many Wharfedale machines are provided with a taking-off apparatus, the most successful and popular contrivance being the simple "flyer." For ordinary book and jobbing work they are to be highly commended, but they are not so desirable when cut, colour, or heavy poster formes are to be done, as the tape and fingers are apt to interfere with the face of the newly printed impression, and sometimes slur.   Neither is the inter-leaving with set-off sheets so conveniently performed when this apparatus is used.

In perfecting, the end laying-on board is removed, the boy standing immediately facing the cylinder, and having a table on his right with the work laid in position.   In ordinary commercial work pointing is sometimes dispensed with, the boy laying to the side mark.   This is placed on the *off-side* of the board, and the sheet carefully pushed over.   Of course it is necessary that the same edge be laid to the fixed point as in the first instance, the sheet having been reversed.   If this is not done, the register is liable to be very faulty, as, had the forme been laid even a thick lead or nonpareil out of the centre of the sheet, the sides would necessarily be out to that extent.   The difficulty of laying-on, especially to the off-side mark, is materially increased if the paper is thin or badly wetted, as it has a tendency to lie in waves, thus rendering dead register impossible.   In this case the points should be used without hesitation, as the spoilage of a few sheets per ream is a serious item.   If the paper is of good substance and comparatively dry, a careful boy should experience no difficulty.

Hard Packing.—This system, universally adopted in America, is now becoming general in this country. There is always to be found a prejudice against any radical change in a method that has been in use for some time.   It was argued that, the material on the cylinder being hard and *unyielding*, the surface of the plates or type would become injured in consequence. But as it is self-evident that a given pressure must be exerted to produce the desired result, it matters little if an elastic material be employed or not. The preparation of the cylinder by this process varies but slightly from that above described. There are several methods employed, only varying, however, in the manner and the number of sheets placed round the cylinder. The paper used should not be heavier than, say, equal to 35 lb or 40 lb double-demy. Take about eight or ten sheets, which should be larger by at least three inches on either side than the forme. Firmly paste these together at one end, and lay and fasten them under the grippers. Allow the pasted ends, however, to be well *inside* the cylinder, as we shall probably have to remove two or three, as afterwards explained. In smoothing the sheets over the surface of the cylinder, take every care that they lie perfectly smooth and flat. For the purpose of underlaying, it is desirable that an extra sheet be fastened over the cylinder sheets to prevent blacking, &c. This operation is of course performed in the ordinary manner, as already described. When everything is ready, *i.e.*, plates tolerably even and colour "run-up," make register, and pull two or three impressions for patching. Then lift the waste sheet and pull an impression on the cylinder. Two sheets, immediately under the one upon which the impression has been pulled, should be removed, as otherwise the impression, when the overlays are put up, will be found to be excessive. It must be remembered that the slightest addition to the face of the overlay "tells," as we are now working *directly upon the surface of the forme*—no spongy blanket coming between. If the underlaying has been well done, thin set-off paper will be found sufficient for our present purpose. Having patched the sheet and cut the pages up separately, by means of a knife cut away any sharp edges, &c., as previously explained; but this ought only to be necessary to the extent of cutting away in one or two sheets, as the cylinder packing will be required to be entirely reserved for the next forme. After the pages are stuck up, take a sheet of paper, lay it flat upon a table, and sponge it entirely over with clean water. Then paste the edges, open the grippers, and lay it over the overlay perfectly flat. Before proceeding further this must be allowed to dry. Sometimes a warm iron is used for this purpose; but it is very much the better plan to allow it to dry without undue application of heat, to prevent contraction. But this, of course, causes a slight delay.

When perfectly dry, run-up colour, and pull another sheet (after putting through a quire of waste). In all probability another sheet will require to be patched. There should be little to do at this stage. Having again pulled the cylinder, it is necessary, before placing the overlay, to remove one or two sheets of the cylinder covering. Carefully raise the whole of the sheets, and detach the superfluous ones—otherwise, with the last overlay, the impression would be too heavy, necessitating the raising of the cylinder. This could be done, but it must be remembered that we have already lowered the cylinder *upon* its bearers, and it would, under the circumstances, be unadvisable to alter the impression screws. Another sheet of paper, damped and pasted as before described, must be placed over the cylinder, and after it is properly dried we are in a position to start. We think it advisable in all cases to put a set-off sheet up, as, in cases of the boy "missing" or failing to place his sheet in time, this can easily be renewed. A dirty cylinder never looks well, and it is better to have a temporary sheet, which can be renewed with but very little trouble. Of course it may be stated that the layer-on should never miss; but, especially in dealing with boys, we should recollect that we must not expect too much.

Another plan, which has been tested and found to answer well, is to dispense with the calico as the *original* covering, this material being used *over* the sheets. As before, place eight or ten sheets round the bare cylinder, and after underlaying pull the impression. Patch, and after putting the overlay in position stretch the calico. This obviates the use of a damped sheet, and inasmuch as the calico is stretched tightly round the cylinder, it renders the covering as firm and unyielding as when the outside paper covering is damped. It is, however, necessary to place a sheet over the overlay. Remove the calico, and pull an impression for the second series of overlays. After this, remove one or two sheets from the cylinder packing, and readjust the calico, when everything is ready for a fair start. In this system there is no necessity to put the damp sheet over the overlay, as the calico can be tightened by means of the ratchet, and the whole will be quite firm, hard, and unyielding. Time is economised, as the printing may be proceeded with immediately the calico is stretched. It is advisable to paste a sheet outside, to prevent the calico from becoming dirty.

When a movable forme is printed off, the whole is lifted. Marks should be made on the coffin at the sides, that the succeeding forme may be placed exactly in the same position, the pitch being determined by the gauge. While speaking of the pitch" or the entry, we may here state that with old machines it will frequently be found that at this particular point the coffin will be slightly lower. This is owing to the jar or sudden jerk when the cylinder is thrown into gear. When this is the case it will be necessary to place two or more sheets on the cylinder, extending as far round as the coffin is low. If a machine-minder is constantly on the same machine (which we may mention is desirable, as he thus becomes used to many singular peculiarities or defects, which every machine possesses in greater or less degree), he will know exactly the amount of extra pressure required at this point. When running the cylinder sheets he will, as a matter of course, provide what is necessary to insure the impression being equalised.

When a stereo or electro forme has been worked off, the plates should be well brushed out with either strong ley or turpentine before being unlocked. After drying them with soft rags, unlock and loosen the catches and lift the plates. The underlay should be carefully removed, and the whole of the pages packed up in signatures. Sometimes the first underlay is allowed to remain, to be used in the succeeding forme, it being argued that the same patching will be equally effective with the next series of plates, owing to the blocks being uneven in parts. We have, however, explained that this is not advisable.

Plates are frequently more injured in lifting and packing than in printing, as there is an inclination on the part of the boys to look upon them as not being worthy of further care when worked off. The machine-minder should therefore superintend their removal and packing, taking the precaution that thick pieces of paper be laid over the face when laid together for removal to the place where they are finally examined, repacked, and labelled.

In the case of colour printing, when there are several

tints, &c., a mahogany block is generally used, as the detached pieces could not be successfully secured by catches. The plates are fastened either by pins or by screws, counter-sunk, the register being made as accurately as possible before the block is fixed in the machine, by means of an impression pulled on an oiled sheet. The plates are laid on the blocks, temporarily fixed, and finally adjusted in their place, the transparent sheet enabling the position to be exactly determined Slight alterations will probably be necessary after underlaying, but if due care has been exercised a little "tapping" subsequently should be all that is required. This will be found fully described in Chapter XXIV.

In the foregoing we have been dealing with the making ready on a single-cylinder machine. The remarks as to patching sheets for both underlays and overlays apply necessarily to all classes of machines, but the manner of proceeding differs in the platen, ordinary perfecting, or Anglo-French machine.

THE TUMBLER.

Although the Wharfedale and Tumbler vary considerably in details of construction, the printer who thoroughly understands one will have but little difficulty in preparing a forme on the other. On some Tumblers of recent construction the double-inking motion has been fitted, adding materially to their value for good bookwork.

When these machines become worn, the cylinder is apt to give the minder some trouble. This is generally owing to the wearing of the brasses, either at the base or at the point of connection of the upright bar with the cylinder crank. It is obvious that if the cylinder shakes

U

either the overlays are thrown off, or a slur takes place. However, the renewal of the necessary brasses is a very slight matter, and as the success of printing depends upon their being tight, they should be periodically attended to.

THE TWO-COLOUR MACHINE.

In laying the formes on the tables, the first forme is fixed at a given distance from the ink table, as in the case of the Wharfedale. But the second is laid not from the edge of the table at the other end, but from the iron bar which marks the division of the coffin.

In a job where two colours are required to fall in the same position, to produce a given effect, it is unadvisable to put it on this class of machine, for two reasons : one is that the first impression, being quite wet with the newly deposited ink, *cannot* possibly take up a further quantity of another shade with any degree of success. If this is attempted it invariably ends in bad work, *as it is impossible to "pick up" on wet ink.* This fact is the real cause of unfortunate failures in several multiple-colour machines—not because the *mechanical* principle adopted was unsound, but that one colour should be perfectly dry before another is put on. We are aware that certain kinds of labels are printed in as many as six colours by one operation. But this work may be described as purely commercial, especially designed in the first instance to meet the objection mentioned above— *i.e.,* by as far as possible allowing each colour to stand by itself. No printer would for one moment profess that its production could be considered good colour work in the ordinary acceptation of the term. If it be desired to print two or more colours at the same time, the tints should fall in different parts of the sheet.

Again, it will be seen that proper overlaying is simply impossible if portions of two formes fall exactly on the same spot, as in the one case a slight impression may be desirable and in the other a heavy "nip." This, of course, may be partly attended to in the under-lay, but sometimes at the risk of rendering the inking defective, as all formes should be quite level, that the rollers may reach every part.

These machines are eminently adapted for two-coloured titles, broadsides, &c., where the colours are defined. The imposition for register in these cases is generally made perfect by the compositor, leaving little for the machine-minder to do when the formes are laid on excepting to see that they are perfectly square.

Underlay plates in the ordinary way, and after the colour is run up—the different tints having been placed in the respective ductors—pull a sheet for patching. On the same sheet of course we obtain the impression of each forme. Patch the overlay and fasten on to the cylinder. If a broadside, the sheet may be pasted up in its entirety, but sometimes it is desirable to cut up into pieces and affix, as in the case of the Wharfedale machine.

The rollers should be used slightly harder than in black work, especially if a heavy tint or body-colour is being worked, otherwise the face will soon become injured.

THE PLATEN.

We have now literally a steam press—frisket, tympan, and a platen affording a perfectly flat impression.

Of course in this machine there is no cylinder to dress, but in lieu of this we have firstly to cover the tympan with parchment, secure the sheets inside, and

U 2

cover the frisket frame with brown paper. With reference to the former, the parchment covering lasts for some considerable time, unless by some unfortunate accident it is cut through by rules, or by a piece of lead or foreign material being left upon the surface of the forme. A frisket should last for some time, supposing that the size of the pages of the formes be the same, and that the position of each set be identical.

The Tympan.—The sheet of parchment for covering should be about two inches larger than the frame all round. Well paste the edges and strain round, tucking-in the extremities on the under-side. With the scissors cut away those parts immediately covering the hooks or pins for securing the top frame in its position. In the latter the smaller extended lips will of course be allowed to be free—these fitting into their slots in the frame of the former. It is desirable that the best or smoothest side of the skin in each case be *outside.* When the paste is set, well sponge the whole, and allow it to dry, which will cause the whole to become taut.

The Frisket frame should be covered with brown paper of good quality. Cartridge is to be preferred, as the former sometimes contains foreign substances, which are apt to become loosened and batter the forme when the initial impression is pulled.

Both the tympan and the frisket should be secured to the knuckle-joints on the outside edge of the coffin, the pin being fastened by a piece of stout copper wire, firmly bent. Prior to putting on the top tympan frame, lay about half a quire of thin sheets of paper on the bottom parchment, and secure them by sewing them with thread at the top end on each side, through the parchment itself. Place the top frame in position, and secure it by means of the small hooks on either side.

In laying-on the forme, see that it is laid *exactly* in the centre of the coffin. If this be not attended to, the platen will have a tendency to tilt, and become strained in consequence of the pressure being exerted unequally The exact position may be easily ascertained by stretching a piece of thread across the coffin from side to side, and from the ink table to the opposite bar. Nicks will be found to mark the central points. Secure the forme in the ordinary way.

We next must pull an impression on the frisket. Roll the forme with a hand-roller, and strike on. When the forme has been pulled, throw up the tympan and frisket, and evenly cut out the impression of the pages on the latter, including about a nonpareil all round, so that there may be no possibility of the sides of the plates coming in contact with the frisket and causing a "bite." The frisket as it stands now is a very frail affair, and would speedily become destroyed or damaged, if not materially strengthened. This is done by fastening lengths of tape (old broken and dirty tapes from other machines are generally used for this purpose) along the backs and gutters on the under-side. Either end is tied to the frame on each side, and fastened to the brown paper by melted composition on the under-side. To prevent any inclination of the frisket to dip, small cubes of cork are glued immediately on the tapes before mentioned. The number and size must be regulated by the dimensions of the paper and the space between them. It is the best plan to cut all pieces at the same time, laying them in their respective positions in the gutters, &c. They should be about a nonpareil higher than the face of the plate, so that when the platen rises the frisket has a slight spring, which tends to prevent slurs, &c. The cork also holds the sheet from off the surface of the forme before the platen comes down, preventing a "double," or dirty mark.

After underlaying the plates, put on the points—one each on the top and bottom side of the frisket frame, and run up colour. The inking on this class of platen is by no means good, and therefore the rollers should receive every attention, that they may be thoroughly fit, otherwise the work will not be first-class. As we mentioned in Chapter VII., an inking cylinder is sometimes fixed parallel with the duct, to assist distribution. It will be seen that practically the platen is held to the beam by a small bolt or socket immediately in the centre, and although it is *guided* by grooves on each side of the frame, it receives no other actual support. Therefore it is sometimes liable to tilt or become unsteady on the impression. If there are blank pages, before pulling for overlaying, place type-high bearers in the empty space. If the bearers are found to be not sufficiently high, a piece of glazeboard may be fastened to the under-side of the frisket, to fall exactly on the top. Of course it is not necessary to cut out the frisket in the case of blanks.

If there are cuts in the forme, lift the inkers and put the overlays, already prepared, face downwards on their respective plates, that the pull may not be deceptive. If this is omitted, when an impression is pulled the patched sheet will certainly be worse in point of impression than before, as the overlay will have borne off pressure at various parts. The placing of the overlay in position is a much simpler matter than on any other machine, as it is unnecessary to cut into single pages. Lift the outer tympan frame, throw back the sheets, removing two or three to allow for the extra sheet, &c., now being fixed, and after slightly pasting portions of the edges, place the sheet face downwards on the points which have pierced the under tympan parchment, and carefully smooth it, so that it may lie perfectly flat. Replace the tympan sheets and readjust the top frame, put in the rollers, run a few waste, and pull a sheet to ascertain the effect of the last overlay. While the second patching is being done, the forme may be washed out, the rollers wiped down, &c., as usual, that everything may be ready for a fair start immediately the next sheet is put up. The tympan sheet should also be affixed. This should be laid upon the table, and pasted at the edges, subsequently damping the whole. It must be then fastened to the face of the tympan, and will become tight and dry before the forme is ready for a start.

We now anticipate that our work has been sufficiently well done to admit of an immediate start being made. Side and bottom lay-marks must be glued on—for white paper—one on the near-side and two at bottom of frisket. These are made by folding pieces of thin glaze-board about three-quarters of an inch wide, bent in concertina fashion. These are of course elastic, and retain their form after the tympan is raised. They must be removed when perfecting. As in the case of the Wharfedale, if the register is slightly defective at one side, by judiciously tapping one of the points the fault may be remedied. A thin sheet should be pasted over the tympan, which can be removed in case of a "miss." When perfecting, this sheet will have to be renewed frequently, especially in case of half-sheet work, when the ink is only partially dry. If the paper has an inclination to crease, cut a few holes in the frisket, to allow the air to escape from between the sheet and brown paper, as possibly this will be found to have been the cause of the trouble.

The platen machine is liable to more accidents in the working than are most other descriptions, and therefore the boys should be most carefully selected. The tympan, with its accessories, may easily become doubled up, if care is not taken in allowing it to fall immediately upon the return of the forme, after the sheet is laid. Another fruitful cause of breakdown arises from the tympan knuckle-joints not being securely fastened. If the wire breaks or comes out, allowing the pin to escape, the side of the tympan and frisket will slip down under the platen, and a break-up is sure to follow. The slide upon which the frames run should also be looked to, as sometimes the bottom screw, securing it to top of side-frame, is liable to work loose.

The laying-on boy should be instructed in the proper method of allowing the table bolt to fall into the slot of the travelling slide, or the latter will strike the pin full, forcing the table back some little distance. The proper time is when the slide is stationary immediately under the pin—when the impression is being taken at the other end.

In making ready, as little impression should be put on as possible, as there is a great strain always on the side-frame at the point where the beam is secured. A break may occur here, if the impression is excessive.

THE NAPIER PLATEN.

As will have been gathered from the description of the platen machines, the Napier is greatly to be preferred to the original. The instructions with reference to the making ready of the forme apply generally to either, but with the Napier the machine-minder will be materially assisted by the additional double-inking facilities, and the sound and solid impression. Perhaps it may be mentioned that it is desirable that the rollers be a little firmer in this instance, as they travel at greater speed, and at least twice the distance. Greater "nip" can be also obtained with an expenditure of less power, by the application of the knuckle motion to the uprights on either side.

THE ORDINARY PERFECTING MACHINE.

The preparation of formes for printing upon this class of machine is in many respects different from that already described. We are now dealing with two formes, inner and outer, and have also a somewhat elaborate system of tapes to watch and keep in order.

As in the case of other machines, it is necessary to stretch the calico over the cylinder as a foundation. Each end is sewn to flat iron bars, these being secured to the cylinder by iron pins. One set of the latter is stationary, the second series being tightened by screws. Upon the calico are pasted about five sheets of paper— say 70 lb. quadruple crown, each sheet being secured to the other by pasting at the ends, as previously described. The calico should be sufficiently large to take the largest sheet the machine can print, that it may be used for any forme. The blanket is stretched over the cylinder sheets and secured by means of pins at either extremity. The above refers, of course, to both inner- and outer-forme cylinders.

If the tapes are already run in the machine, the cylinder sheets, in the case of the outer forme, are first placed in position at the point above the delivery. The cylinder is stopped at such a point as will allow the front edge of the impression portion to be placed and the paper to be fixed. The machine must be slowly moved forwards, that the whole may be smoothed down. At the position referred to there are no tapes to interfere with the adjusting. The whole series of sheets should be rolled over or doubled, so as to occupy but little space, and pushed through from the near-side. By reaching over the taking-off board, the front edge may be easily fastened by paste, and the boy, slowly turning the rigger, or cylinder wheel, moves the cylinder so as to allow the rear end of the sheets to be secured. The blanket may now be placed on in the same way, and pinned, taking care that it is perfectly taut.

With the *inner* forme, the cylinder sheets and blanket are first placed in at the point between the gripper or drop-bar drum and the first reversing drum, as here, again, no tapes interfere. All that is necessary is that the machine, as before, should be moved to such a point that the fore edge may be in correct position, and, when the sheets are secured, moved forward slowly to allow the workman to smooth the sheets or blanket. In this case the machine-minder generally kneels upon the laying-on board facing the outer forme.

Prior to laying on the first forme of a job, the machine must be taped. Fig. 95 shows the travel of the sheet, and also indicates the run of both the inner- and outer-forme tapes. Take a reel of tape, and first put it round the inner-forme cylinder, over the first drum, under the second, and finally over the outer cylinder. Stretch it round the wooden roller immediately in front of the taking-off board, and take it down under a similar roller beneath; carry it parallel under the cylinder, out-side the roller, and thence over the tightening pulleys and above the drop-bar or the gripper-drum, sewing the ends together. The whole should be tolerably taut, as it will be sure to slightly stretch at first in the working, when it can be tightened by the tape pulley above the outer forme. Having run one tape through the machine in the manner described, the next may be pinned to it say at the outer-forme end, and the machine be allowed to run very slowly until the end reappears. Cut it off from the reel, and sew as before,

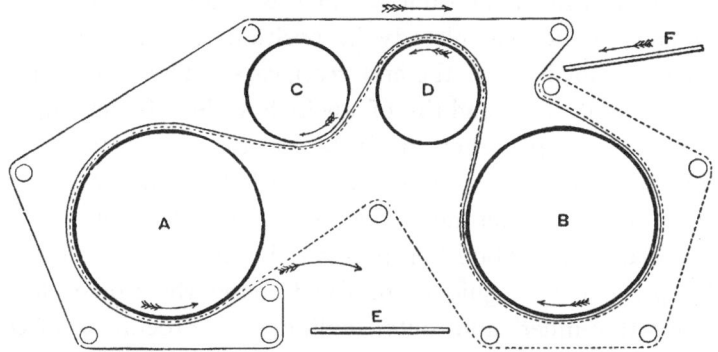

Fig. 95.—Showing Travel of Tapes on the ordinary Perfecting Machine.

A Outer-forme cylinder ; B Inner-forme cylinder ; C D Intermediate drums (D Register drum) ; E Taking-off board ; F Laying-on board.

when it can easily be shifted in position by slightly pulling one point towards you, making it a fixed point, as it were, and running the machine slowly. The remainder of the inner cylinder tapes may be run-in in a similar manner ; the number, of course, being determined by the pages, one set being required to run between each set of gutters. Having finished the inner, we pass the first *outer* tape in.

In running-in the outer-forme tapes, start from the tightening pulley over the outer-forme table, over both the roller above the register drum and the one on top of the gripper or drop-bar drum, continuing round the latter and the inner-forme cylinder, over and under the first and register drum respectively. From this point it is taken over the outer-forme cylinder, above the delivery roller and down under the bar immediately below, and then horizontally to a companion roller near the base of the cylinder, and joined at the point where we commenced. The remainder of this second series of tapes may be run-in in the same manner as before, by attaching an end to the one already in position, and running the machine slowly until the end reappears. The tape generally used is three-eighths of an inch wide. In arranging the sets of tapes they must be so guided as to travel exactly together—*i.e.*, one set immediately on the top of the other. Sometimes they will "wander," or travel irregularly, and "bite" or pull on the edge of the forme. To prevent this, guides are fixed round the iron rollers, under the inner and outer cylinders. Procure a piece of tape, compo one side, and attach the end to the roller, and turn it, pressing the tape well. The sides of the guides thus made will keep the running tapes exactly in position.

Obtain, if possible, a pull, however rough, of the forme which is about to be put on, when by laying it flat on the outer-forme coffin, the tapes may be arranged so as to render subsequent alteration unnecessary. Place the forme in position, using the gauge for determining the "pitch," measuring as usual from the edge of the ink table. When locked up in the coffin, move the machine slightly, that the end of the forme may be in close proximity to the cylinder. The correct position of the tapes may now be determined to a nicety. They should in every case be immediately in the centre of the gutters, and can be easily shifted by slightly altering the guide pulleys in the desired direction. Having so arranged the tapes as to insure a free run, the inner forme may be put on the other coffin. Place a waste sheet over the surface of the plates to prevent soiling the blanket. Having arranged the pitch, again measuring from the ink table, the position for centreing may be ascertained by the outer-forme tapes, already adjusted. It is, however, advisable to measure the exact distance from the edge of the cylinder bearers, and compare with the outer, before locking up, when but a slight alteration of position will be necessary in making register.

The impression should be next adjusted. Run the outer forme out, lay on the surface a clean sheet, and move the machine slowly, allowing the tables to travel in and out, when it will soon be seen if it is desirable to alter the height of the cylinders. The impression should be comparatively slight—in fact, just enough to show the heaviest portion of the plates plainly. If too strong, slightly raise the cylinder by loosening the top impression screw and tightening the lower. If not sufficiently firm, reverse the process. In manipulating these impression screws, be sure that they are subsequently well tightened, or the impression will be deceptive. Unless absolutely necessary, as before mentioned, avoid altering the impression screws.

Roll the forme with a hand-roller, and put a sheet over the surface and allow the coffin to run in and out. Patch the sheet, and paste the underlay under the plates, in manner before described. After planing, and slightly tightening the quoins, run the machine a few times, with waste sheet on forme, and then roll again and pull a clean impression. After patching another underlay, the inner forme may be treated in the same manner.

The packing should be placed on cylinder bearers, as before mentioned (p. 291), of both inner and outer formes.

The vibrator, wavers, and inkers, must be trimmed, &c
As described on p. 292, the vibrators may be cut to suit
the forme, and the wavers and inkers pared at the ends.
After putting the ink in the ductor and adjusting the
metal stops, to regulate the flow of ink, have the wavers
and inkers sponged down.   Run the machine until the
ink table is partially covered, and place the wavers in
their respective forks, two at one angle and two at
another, and allow the machine to run until the ink is
fairly distributed.   A waste sheet should be placed on
the face of each forme while this is done, and it is
desirable that a boy be at either end to keep it flat,
or the wavers are apt to catch a loose corner, and tear
it up.   When sufficient ink has been deposited, run two
sheets of waste in from the laying-on board, lift the
waste sheet from the forme, and place the inkers in
position.   Run a quire of waste through, and pull two
or three good impressions—one of which should be sent
to the reader for revision.

The register should now be attended to.   It is
desirable that the slight alterations that may be necessary should be made, if possible, in the outer forme, as
this is much easier of access.   When any corrections
marked in the revise have been made, and the locking
up finished, the sheets may be pulled for overlaying.
Run the waste.   Stop the machine when the outer forme
is out, lift the inkers, and place a clean sheet on the
plates and pull an impression.   Lift the inner-forme
inkers and also pull a sheet, when the patching of
the two formes may be proceeded with.   If there are
cuts in the work, the overlays, if already prepared,
should be placed face downwards on their respective
places, that the exact pressure may be apparent.   It is
convenient now to pull the cylinders ; the blankets must

be partially removed. Before unpinning, mark either
side with chalk, that they may be replaced exactly in
the same position. Remove the pins from the under-
side, and, in the case of the inner forme, lift the end
at the point just below the receiving drum, allowing
the impression sheets to be bared. In the outer forme,
the blankets may be turned back at the point above
the taking-off board. It will be seen that the blanket,
now thrown back, secured at the same time at one
edge, lies over that part of the cylinder not utilised for
the impression. By striking-on the machine gently, an
impression is pulled upon the cylinder sheets.

The overlay may now be patched, the boys in the
meantime cleaning up formes, tables, rollers, &c. We
have not in the foregoing stated the average time that
the patching of a sheet for overlay or underlay should
occupy. This is obviously impossible, as the amount
of work to be done depends entirely upon the dimensions of the forme, the number of pages, and also upon
the state of the plates and the character of the work.
When the patching is completed, cut the pages up
separately, and paste in their respective positions upon
the cylinder sheet, when the blanket may be turned
back to its original position, secured by the pins.

Run up colour, and, presuming another overlay is required, proceed as before. After pasting up the second
overlay, we make preparations for starting. The set-
off sheets must be placed in position. These consist
of narrow slips a little wider than the pages, and
long enough to extend round the outer-forme cylinder
and over a bar immediately above the delivery. These
should be cut into lengths, allowing about six inches
lap for pasting. In order that they may travel exactly
straight, thick indiarubber rings are placed round the

set-off roller, and these can be shifted into any desired position. It will be noted that inasmuch as the
length of the strip is greater than the circumference
of the cylinder, in consequence of its travelling over
the bar mentioned, it changes its position at every impression. This is a material advantage, as consequently
it is not necessary to renew the strips so often. In join-
ing, let the paste be as thin as possible, or the mark of
the join will be apparent in the impression.

The rollers, inkers, and wavers having been wiped
down, run up colour as before, first with wavers only,
and after running two sheets into the machine, put
the inkers into position. Feed in a quire of waste,
and pull say half a dozen on "its own." If any defect
now exists, it will possibly be in the inking, in which
case the ductor screws may be carefully altered. If
the register is not perfect as a whole, a turn of the
bar above the registering drum (the one next the outer-
forme cylinder) may effect the desired result.

A fair start having been made, sheets should be
continually removed from the taking-off board during
the running, and examined. The taking-off boy in this
case cannot always call the workman's attention to any
defect in the printing, as the sheet is delivered almost
in the dark, and as the machine runs at a tolerably
quick speed, his whole anxiety is to lay the sheets
properly as they are thrown out. It will be noticed
that the drier the paper is, the easier the taking-off
becomes, as the sheet is thrown out almost flat upon
the board ; whereas, if too damp, it possesses a tendency
to double up immediately the front edge is delivered
—often rendering a stoppage necessary. A bank is
generally placed by the side of the machine, upon
which the sheets may be spread for examination.

With reference to the set-off bands on this machine, the time they continue in good condition depends greatly upon the nature of the forme being printed. If there are many or heavy blocks, they will soon become soiled, necessitating renewal. If the paper employed is too soft, they are apt to tear at the edges, and will soon be destroyed. When it is necessary to run one in, it may be done in a somewhat similar manner to that adopted when renewing a tape. In the first place, the piece of paper should be rolled up in single lengths ready for immediate use. When a new one has to be substituted, stop the machine with the inner forme out. This will enable the slips to be passed round the outer cylinder without difficulty. Paste the end on the one to be supplemented, tear the latter across *below* the pasted end, and pull it towards you until the first portion of the new piece comes to hand. Release the pasted portion, remove the old slip altogether, and join the new slip as before. Oiled and prepared papers are frequently used for these continuous set-off bands, which last very much longer than ordinary plain paper, as they are not liable to become soiled so quickly, besides which they are more tough.

When a series of formes is being printed at the same machine, there is, of course, less work to be done in the preparation of the second and following ones, as the stereo blocks are already placed, and the tapes are in proper position.

It is advisable to examine the quoins of the forme at intervals—when fresh paper is being put upon the board. The reversal of the tables, as explained in Chapter VIII., is somewhat sudden, and the jerk is liable to loosen the locking-up, causing the furniture and catches to rise, which produce unseemly

V

blacks. It will also be found necessary, especially when the number is long, to wipe out the coffins two or three times a day, as the dust or fluff from the paper accumulates, and is picked up by the inkers and deposited on the forme, causing picks. These remarks more especially apply to this class of machine, on which, as a rule, the paper used is of a commoner quality than that adopted for better work on single-cylinder and Anglo-French machines

The construction of the perfecting machine is by no means of a complicated nature, and with common care in greasing, oiling, &c., it should run for years without requiring any costly repairs, accidents of course excepted. The pinion-wheel and rack should be well greased before starting, and the impression pulleys be oiled, that they may run freely, otherwise, as is the case with all wheels of this description, " flats " are likely to occur. The runners should also be well lubricated, that the table bars may have an easy travel. If it be found that any of them are low and not flush with the bars, the brass bearings may be slightly packed from the underneath to bring them to the required height.

The tape pulleys and bars must also receive occasional attention. The tapes should always have a free passage, and not be allowed to drag. When they become too loose, tighten them by means of the pulleys over the inner or outer formes, as the case may be. Never allow a dirty tape to remain in the machine. They are often more soiled by the indiscriminate use of dirty waste than in legitimate working. Although for reasons of economy old sheets may be used for waste, when they become black and limp another lot should be substituted, or the tapes and blankets will quickly become blackened

THE ANGLO-FRENCH MACHINE.

Necessarily many of the instructions for making ready on the ordinary perfecting machine apply also to the Anglo-French. Usually, however, the kind of work selected for this machine is of a better class. It travels slower, the inking is superior, and the arrangements for preventing set-off are certainly more effective. The cost is proportionately greater, as, besides the output being about one-third less, four boys are employed—two layers-on and two takers-off being necessary. As already mentioned, these machines are sometimes fitted with the upright spindle to enable them to run at a higher speed; but the amount done must necessarily be regulated by the capability of the boy in lifting the set-off from the printed sheet, and it is practically impossible for the boy to take say a thin quadruple-crown sheet and lay it squarely on his heap at a greater speed than 1,000 per hour. If the speed be increased, the single set-off must be dispensed with, and we might almost as well adopt the ordinary fast perfecting machine. It is true that the plan of continuous set-off sheets has been at various times attempted, by fixing a roller above the outer-forme cylinder, over which the bands travel; but in no case have they been pronounced an absolute success, partially owing to the grippers, which are apt to tear the paper in their opening and closing.

Unlike the machine previously mentioned, the Anglo-French possesses only two cylinders, and therefore the taping is not such a difficult matter.

First cover the cylinder with calico. Two rods are provided—sew the material on to the flat one, and drop it on to the small pins inside the cylinder under the grippers. Put it round, and fasten the other extremity

V 2

to the ratchet-bar, when it may be made perfectly taut. Over this lay about five sheets, fastening under the grippers, and, allowing the machine to be moved forward, smooth them well over the cylinder in the ordinary way. The blanket is secured by an iron bar, holes being made in the blanket to allow it to pass through at short intervals. This bar is lodged behind the studs under the gripper, and stretched tightly round the cylinder and pinned. This applies equally to both inner and outer.

As in the case of the ordinary perfecting machine, the position of the tapes should—if possible, in the initial forme—be determined with the assistance of a pull from the forme. If this is not available, a sheet can be easily folded, by which means the position of the tapes may be temporarily arranged.

The grippers in each cylinder should be looked to and finally set; and, as those of the outer forme enter the inner-forme cylinder, care must be taken that they are fixed firmly, and in such a position as not to touch, otherwise a serious break will occur. The inner-forme grippers should be placed near the extremity of the sheet at either side, so as to insure its being taken accurately when laid to the marks.

Lay the inner-forme in position, the pitch being decided by the gauge—measuring from the inking table. The tapes will of course be a guide as to the position of the outer-forme, care being taken, however, that exactly the same space be allowed between the inking table as in the case of the inner-forme. Sometimes the pitch of the formes varies, necessitating a gauge to be kept for either end. Roll the plates with a hand-roller, and pull an impression for the underlay patching, as before described.

On all gripper machines, in setting the grippers, judgment must be exercised in determining the amount of pressure to be exerted. The grippers, as before explained, are arranged along a bar, and are entirely dependent upon the said bar for their motion. When adjusting, the set-screw should be loosened, and the gripper pressed firmly to the edge of the cylinder and secured. When the entire set has been adjusted and fastened, test each one by placing underneath a piece of paper, and by pulling the paper away the amount of pressure exerted may be easily gauged. Care must be taken that the grip is not too tight, or a series of marks will be left on the edge of the sheet, which the hydraulic press sometimes fails to remove. On the other hand, unless the paper be held perfectly tight, the sheet is liable to be torn away in the printing.

When the underlaying has been completed, pack both the cylinder and roller-bearers in the ordinary manner, prior to inking up and making register. After adjusting the ductor and trimming the vibrator, place the wavers in their respective forks—taking care that the foremost ones in both inner and outer forme have the movable lift adjusted, otherwise this waver will touch the surface of the foremost pages of the forme. This machine has a longer travel than the ordinary perfecting machine—and the front waver really runs beyond the entry—necessitating the lift as mentioned above. Sheets will of course be allowed to remain on the forme while running-up colour, before the placing of the inkers, prior to which one sheet is fed-in from the laying-on board, and the machine is stopped just as No. 2 sheet is secured by the grippers, when the inner-forme will be immediately under the laying-on board. Place the inkers in the forks, and run a quire of waste, taking care, however, to have the set-off sheets fed-in also in the ordinary way, to prevent the soiling of the outer-forme cylinder blanket.

When register is perfect, and the revise has been attended to, impressions of both inner and outer forme may be pulled for overlaying, the inkers having, as before, first been lifted. Overlays for cuts should be placed downwards on their respective blocks, as usual. The cylinders may now be pulled. Unpin the blankets, after marking at the sides to insure exact readjustment, and *entirely remove them from the cylinder*. Allow the machine to run once, and have the rollers, plates, &c., carefully cleaned. When finished, cut into pages and paste on cylinder. While the minder is doing this a boy should be stationed at the fly-wheel, to move the machine in any desired position. When " backing-up," however, the ductor cord should be loosened, or the roller will be reversed, and the ink emptied in bulk on to the table. It may be advisable to remove one or two sheets from the cylinder, as the overlay will obviously increase the impression, and it is better to lessen the pressure by this means than to alter the impression screws, unless the overlays are very thick. The blankets should now be replaced, and carefully stretched to the position they formerly occupied. When all is clear, place a sheet on both inner and outer forme, and allow the machine to run for two or three minutes, that the overlays may adapt themselves thoroughly to the plates. Have the rollers sponged, and colour run-up, everything being now arranged for a proper start, although another overlay·*may* be required. It is, however, desirable that we ascertain the true state of the work by properly inking, that everything necessary may be done to the second and presumably the final overlay. After running, say, a quire of waste, pull about two or three of "its own." The general "go" of the work may be now ascertained, and any defects in the inking rectified. It may be necessary to trim the vibrator, or to allow more ink to flow at various points, the sheet showing plainly what is necessary in this respect. Having readjusted the inking, run waste as before, and pull half a dozen clean sheets, (feeding-in set-off), and then, if the general appearance is satisfactory, pull the sheet for patching. The machine-minder should himself lay a sheet to the mark and strike-on the machine, stopping it just before the inner grippers open to release it into the outer forme. A boy should be stationed at the fly-wheel to hold it *immediately* the strap is struck-off by the minder. Lift the grippers and carefully remove the sheet, which of course is only printed on one side (the inner forme). Substitute another clean sheet, strike the machine on full speed (taking care to have a waste ready laid to the marks), and the sheet having only the impression of the outer will be thrown out on to the taking-off board. Be sure, however, to have a set-off stroked in, in the ordinary way, that the impression may be thoroughly fair.

Notwithstanding that we have already stuck up the carefully prepared overlay, it is very likely that, in comparing the impressions with the India proof, the solids will require to be strengthened and the lights cut away or softened. Several sheets having been pulled, these impressions may be used for the purpose. After patching, remove the blanket again, and paste the over-lay in position, and we should now be ready to make a fair start. Another overlay yet may be required, but only in exceptional cases should it be absolutely necessary. It should be the constant aim of a machine minder to accomplish all that is necessary in two under-lays and two overlays, and we are sure that it will be generally admitted that if, in each of these stages, the work be thoroughly and discreetly done, a third overlay will not be required. As before stated, in some instances it may be desirable ; but we are now dealing with the ordinary run of book-work. We are aware that some minders are dissatisfied with the result until the *third* overlay is put up, simply because they have always done it. This is, however, more often a matter of habit than of necessity.

After having finally fixed the blankets, have the forme well washed out and the wavers cleaned, and run-up colour. Competent machine-minders often prefer to sponge the inkers themselves—a practice to be commended, for the success of the work largely depends upon the condition of the rollers. Being satisfied that these are fit and sharp, run two sheets in the machine, and place the inkers in their respective forks. The first ream of white paper will in the meantime have been placed on the board, the long slips of cardboard hanging from the front edge of the laying-on board having been marked for exact lay, and the set-off sheets placed in position. After a quire of waste has been run through, pull a clean sheet, when, if any defect still presents itself, it will possibly be in the inking. This having been remedied, run waste finally, and make a fair start.

As will have been noted (Chapter IX.), this machine is somewhat complicated in construction, and therefore it is absolutely necessary that every precaution be taken that each part is frequently oiled or greased, otherwise undue wear will occur in some portion, resulting in endless difficulties. Grease should be used for the rack and pinion wheel. The *segments* at the end of the rack, together with the tumbler, must be examined and well lubricated, as in one case the steel surface is liable to be cut, and the tumbler to be worn unevenly. When the tumbler is travelling round, it should be flush to the segment, otherwise a jar will occur, and too much play be allowed on the end tooth. With a thorough knowledge of the construction, however, a very little experience will enable the workman to become acquainted with those parts requiring extra attention.

With reference to the set-off sheets, it is really no economy to employ very thin paper, or to use it too often. The boy should be impressed with the fact that although a slightly bad lay may not be a very serious matter, it is necessary that care be used in the stroking-in. It is a common error to suppose that the stroking-in of the set-off is a very secondary matter, for want of care may result in two or even more sheets being taken, which, it is hardly necessary to say, affects the impression of the outer forme. Again, the taker-off should be instructed to put aside all torn sheets as unfit for future use, as they are liable to become destroyed in the tapes, and, dropping upon the rollers, necessitate a stoppage.

MAKING READY—GENERAL REMARKS.

When a forme is received from the composing room, it is presumably in a fit and proper condition. But the machine-minder knows from experience that, like himself, compositors are not infallible. But, given that every forme is well gauged, it may possibly be found that the register is slightly defective, even in the case of movable formes, and slight alterations are therefore necessary to insure perfect fit. Bad register is inexcusable. It is perhaps worse than bad printing. A page out of the centre is apparent even to those who are unable to detect defective printing. Even with the commonest class of work every forme should be *absolutely* registered before overlaying, leaving nothing to be ultimately adjusted.

The aim of a printer should be to produce *clean* work. Work full of ink never looks well in bulk, as the density of colour distresses the eye, besides imparting a general smutty appearance. If some of the volumes printed, say, thirty or forty years ago, are examined, it will be found that they possess a delicacy which is sadly wanting in many more recent kindred works. We are aware that they were printed at hand-presses, but the pressmen seemed to understand the fact that, without making the work positively grey, it was certainly undesirable to overload the forme with colour. With automatic rolling, it should not be difficult to insure accurate deposition of the ink, as the facilities now provided give the workmen, with ordinary care, almost unlimited control over the rolling ; and if the sheets are constantly watched after a good and legitimate start has been made, little trouble should be experienced in obtaining an absolutely satisfactory result. Machine-minders should never make it a principle to leave it to their boys to call attention to anything radically wrong in the way of colour, &c. If even a quire of work be fuller or lighter than the bulk—owing to the colour having unexpectedly run up, or down, it must be remembered that practically as many volumes as there are defective sheets will be marred. With cut-work it is found necessary to work the colour darker, as of necessity the engravings will require well filling. But presuming the making ready to have been well done, the colour should not be allowed to run too freely. Work that is too full of colour is suggestive of slovenly making ready.

The difficulties of the machine-minder are necessarily increased when the type or plates are old. Sharpness is impossible when the face of the forme is worn, and after obtaining a spare impression the colour should be comparatively grey. In this case we would not advise hard packing, as a spongy impression, so long as it is perfectly accurate, is less liable to expose defects than an unyielding surface.

Slurs are frequently caused by placing an improper quantity of sheets round the cylinder. The surface of the latter must of necessity travel exactly with the face of the forme If the diameter of the cylinder be increased by pasting up an excess of sheets, it will be readily understood that the two planes must of necessity travel at differing speeds—causing a slur, or smudge, as before stated. Another fruitful source of slurs is general looseness. The table should be held rigidly between the side-frames, and the cylinder bearers be closed firmly round the shaft. If there be a shake in any part of the machine, owing to wear in the bearings or bad lubrication, a jar will result, causing a slur in the printing.

Friars are usually the result either of bad packing of the rollers, or of some foreign material having been deposited on the bearers, causing the roller-wheels to jump, thus preventing the deposit of ink on certain portions of the forme. The packing must not be too sudden at either extremity, but should be carefully pared down on each side. This is necessary also to prevent the roller from jumping. Always examine the bearers, and take out the tacks that have been used to secure the packing used for the previous forme. When the bearers are badly worn, it is the best plan to have them renewed.

They are comparatively inexpensive, and when in good condition save time in both making ready and printing.

Monks are frequently caused by the bad condition of the ductor-knife allowing the ink to escape in thick ridges, which the wavers are unable to distribute before the inkers take it up. The only remedy for this is to have the knife re-ground. The vibrator may be cut at the point, which will, to a certain extent, obviate the defect, but this is a clumsy expedient. Again, monks are sometimes caused by the marks of the mould on the roller. Before placing in the machine, the rim of composition marking the join in the mould must be carefully pared with a sharp knife, or the ink will accumulate, and become deposited across the surface of the forme. Rollers the composition of which is perished will also cause an uneven deposition. If the surface be leathery and hard, it is impossible to obtain good rolling, and if after sponging-up with lukewarm water the face speedily relapses into the same state, the roller should be discarded.

Riders are supplementary rollers, or iron rods, placed upon and immediately above the inkers, the object being to assist the deposition of the ink upon the forme by giving additional weight. They are not so much used on the Wharfedale machines as formerly, owing to the increased number of inkers provided. Besides this, the stocks are now made of greater weight, which renders the use of the rider unnecessary.

There can be little doubt that the rider is apt to affect the face of the composition, not so much because of the weight on the top, but that they are liable to jump and skid, which rubs the surface of the roller underneath. Therefore, except in the case of very limited inking they should not be used. Various arrangements

are adopted in order to prevent their jumping out of the
forks. Holes are sometimes drilled through the tops of
the forks, and a piece of stout iron wire pushed through,
thus preventing the possibility of their becoming dis-
arranged. Some makers fix a movable iron band, working
on a pin at each extremity of the forks, enabling the top
of the slots to be safely closed. On fast machines, when
riders are used, some safety arrangements should be
adopted, as serious breaks may occur if the rider happens
to be jerked out of position.

NOBLE'S SPRING ADJUSTMENT

renders the use of riders unnecessary. It consists of
a small slide, which may be fixed by a nut to the

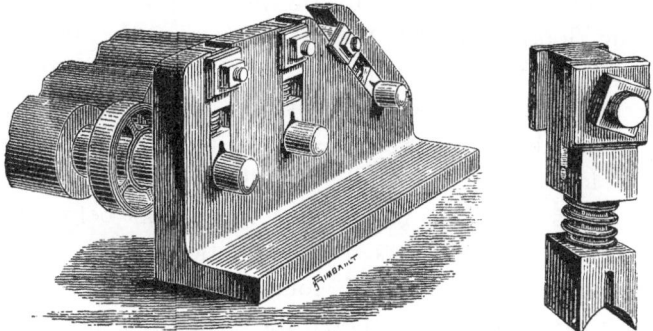

Fig. 96.—Noble's Spring Adjustment for Inkers.

slot of the fork. Immediately underneath is a spring
attached to a small semicircular shape, which presses
upon the spindle of the inker. By this means the
rollers may be made to travel upon the surface of
the forme at the desired pressure, thus obviating the
use of riders. It will be understood that this really
approaches the system adopted on litho machines, on

which springs are attached to the roller spindle, and fixed to a hook attached to the side-frame.

Care is necessary in the use of this adjustment, as, if too great pressure is exerted, the face of the roller will speedily become injured by the sharp edges of the forme. However, we think that, with a due amount of caution, its employment will be found of considerable assistance, especially with heavy cut-work.

Warped Blocks.—When working movable formes in which are interspersed blocks, frequent troubles are caused by the latter showing a tendency to warp. When the block consists of an electro or stereo, mounted, there should be but little difficulty from this cause, supposing that well-seasoned mahogany has been used. Frequently, however, cheaper wood is utilised, and then the wood will shrink from the plate, and the pins will become loose. When this is the case, the best and quickest plan is to remount the plates upon a piece of reliable material, as otherwise not only will the printing of the block itself be defective, but the appearance of the letterpress surrounding it will certainly be affected also. It is advisable to examine the blocks before commencing to underlay, as a few additional pins may be required. Care should of course be exercised in the driving, and the head should be sent well home with the assistance of a punch, or it may rise and black.

Boxwood blocks will often warp, which may sometimes be attributed to their being allowed to become over-damp with ley. Turps should always be used for cleaning, and in very small quantity. After brushing out, well dry the surface with a soft, clean rag.

If a block is found to rock, it should be immediately taken out of the forme and placed face downwards upon a flat surface with a piece of soft thick paper between. Put a weight on the back, and allow it to remain until it becomes thoroughly level. Care must be taken that too much pressure is not exerted, otherwise the wood will break, or the joints will come apart. When this latter is the case, have it locked up tightly in a small chase, moulded, and an electro taken. The cracks in a plate may be filled in and touched up, but it is impossible to "tinker" up a wood block, which is sure to become worse if not attended to.

Blacks in movable or plate formes are mostly attributable to bad locking up. The vibration of the machine, together with the lug of the rollers, quickly loosens the spaces, furniture, or catches, and they become decidedly troublesome. It should therefore be a special care, after the making register has been completed, to take every precaution that every quoin is well driven home. It is not necessary, however, to make them too tight, or the pages will spring, and cause the very trouble we are referring to.

Bad justification in movable, and slovenly making up of stereo blocks, are sometimes the cause of blacks. This is owing to downright bad workmanship, and no amount of judicious locking up will keep the spaces, furniture, &c., in their proper places. There is, of course, no remedy for this. In an office where such work is allowed, good printing is impossible.

Fixing the chase too tightly in the coffin will cause it to rise, as will also careless placing of the furniture forming the wedges on each side. A very slight experience, however, will soon dictate the safest and best manner of securing the chase in position. The furniture should be carefully selected, and thoroughly solid. When a small forme is placed in a comparatively large bed, iron chases may be advantageously used to fill up the space.

CHAPTER XXI.

ON THE CUTTING OF THE OVERLAY.

A PATCHED sheet placed round the cylinder of a Wharfedale or a perfecting machine, or inside the tympan of
the platen machine, is termed, generally, an "overlay"—in
distinction from the "underlay," or piece of paper, put
under the blocks or plates. We have already dealt with
both the above, and now propose to describe the method
of preparing the overlay for cuts.

It must be admitted that complete success is largely
dependent upon the engraving itself. Some engravings
are cut in sharp, distinct lines, bold and effective, while
others are close, and consequently shallow. In one case
the labour is comparatively slight, a level impression
often yielding fairly satisfactory results, while in the other
the work is materially increased, and the ultimate effect
often unsatisfactory.

Since the proper making of overlays takes some little
time, it is advisable to have them prepared prior to
laying the forme on machine, so as to save unnecessary
delay. Impressions should be pulled at hand-press from
the blocks or plates to be printed from. It is perhaps
unadvisable to take the pulls from a wood block for this
purpose, if it is to be subsequently electrotyped for machine,
as the plates are frequently of slightly different size from
the original, owing to the uneven contraction of wax in
the moulding. Added to this, blocks will frequently warp,
and sometimes split when under pressure.

The thickness of paper used must be to some extent

regulated by the system of making ready to be adopted —with or without a blanket. It is patent that the effect will be much more apparent when the overlay comes into direct contact with the cut—*i.e.*, with only the thin cylinder-sheet intervening—than when a thick woollen blanket is employed. Really delicate effects can rarely be produced in the latter case, as the blanket is apt to interfere with the true definitions of sharp lights and shades.

Presuming we are cutting an overlay for a cylinder hard-packed, the electro should be laid upon an iron bed in the press and (say) five impressions be pulled. It will be found sometimes necessary to slightly underlay the cut to obtain a perfectly level impression. The pulling of pieces for overlaying is very often looked upon as a very unimportant matter; but if ordinary care is not exercised, the work of the cutter will be materially increased. One of these pieces should be printed on tolerably thick plate paper, hard paper being used for the remainder. The ink must be stiff, and care must be taken that the plate is well and thoroughly rolled, so that the impression may be as perfect as possible. Avoid the use of too much ink, or some of the half-tints will be too dark. If this is not attended to, it is apt to create an uncertainty in the manipulation, and render constant reference to the India proof necessary. Before commencing to cut out, allow the impressions to become thoroughly dry.

The object of an overlay is not to equalise the impression, but to intensify the pressure upon the dark parts or solids, that they may be firm and bright, and to lessen the impression upon the lighter shades, in order to give them that degree of delicacy and cleanliness that would be altogether wanting if the pressure exerted were uniform. Supposing an engraving to possess prominent solids, together with graduated tints, were the impression

W

sufficient to give the necessary depth to the blacks, the tints would appear dirty and harsh, and almost perforate the paper, owing to the distinct and sharp lines. On the other hand, if the pressure were regulated to suit the tints, the solids would appear rotten. This can be easily understood, when we consider that the fine lines are sharp and piercing, and sometimes necessitate the use of a bearer to prevent the paper being forced down on either side of the thin line, while the solids in themselves are capable of resisting a considerable pressure without injury.

Before giving examples, we may say that generally an engraving possesses five different gradations, or tones of colour—the extreme solid, the semi-solid, the half-tint and the open-line work, and finally the lightest tints.

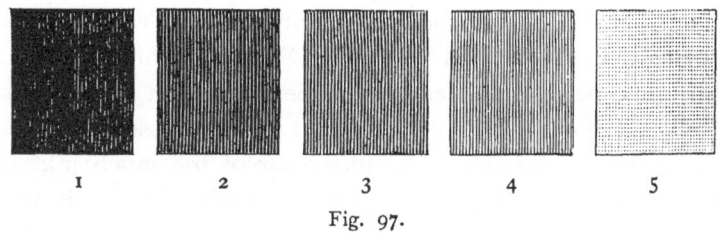

Fig. 97.

The above will give a general idea of what is meant, it being understood that the example has been cut, not as a sample of engraving work, but rather of colour.

First procure the India proof of the engraving. This is, or should be, perfect. It is the engraver's impression of his own work ; and as it is often passed by the artist who made the drawing, we may assume it to be the standard of excellence. It is perhaps as well to bear in mind, however, that the means used in its production are very different to those employed by the printer. Superfine ink and paper are used, and by the aid of a burnisher, and by "wiping," effects are frequently produced which it is impossible to equal by mechanical means. However, it is far better to have a super-excellent proof—with effects exaggerated—than to be left to our own idea as to how a cut should appear. It must be the aim of the overlay-cutter, by so adjusting the impression, to follow as clearly as possible the general "go" of the India proof.

The pulls being perfectly dry, take the impression printed upon the thickest piece of paper, and cut out the "whites," or those portions upon which there is absolutely no work. The high lights, or palest tints, should be peeled. Run the knife lightly round the parts adjacent to the middle lights, and carefully peel or scrape a layer of paper away. In no case should the paper be cut sharp down. It is advisable to pare the edges, otherwise the distinction in the impression may be too sudden or marked. After this has been done, take one of the pulls on thin paper and cut out the solids, or blackest parts, and with very thin paste place the pieces *exactly* in position on the first pull. The third impression must be more general, containing the tints marked 1, 2, and 3—in fact, embracing all the forcible parts, cutting out the lighter tints, 4 and 5. From the fourth or last impression only the lightest tones should be cut out (5). Rarely are five pieces used, as the last would require to have the high lights omitted ; and if, when the overlay is finally placed in position, it is found that the lightest tints suffer, a piece over the entire cut will generally effect what is required.

It must be remembered that the cutting of an overlay is not merely a mechanical operation. Unfortunately, it is frequently looked upon as such. Although every machine-minder should be capable of bringing up a cut in an

W 2

artistic manner, we are afraid that such is not always the case. To turn a sheet to the light and determine where a patch is required to equalise impression is a very different matter from ascertaining if the appearance of a cut answers the expectation of the artist, engraver, or, in fact, the critical public. The quality of printing, at the present time, is advancing rapidly, and the "general reader" is able to discriminate between honest and careless printing, especially in the case of "pictures."

If really good printing be required, the overlays should never be scamped, otherwise endless trouble will be the result; for when an impression is pulled, it will be found thoroughly unsatisfactory, and another overlay will have to be put up, thicker than should be necessary, to remedy in some measure the defects consequent upon the first being improperly done. This, of course, affects the general impression, as the additional thick overlay bears the pressure off the letterpress immediately adjoining. Then commences a plentiful distribution of patches on the parts affected—resulting generally, in the long run, in loss of time and unsound making-ready.

In cutting overlays for machines whereon blankets are adopted, it is desirable that slightly thicker paper be used, as the effect of the various pieces is not nearly so marked as when only a single sheet covers the cylinder.

Overlays should be as thin as is consistent with securing the desired result. If the impression on the various tints has been so adjusted as to produce an artistic effect, it becomes simply a matter of dead impression, which may be dealt with by placing a piece of paper over the whole. In dealing with the high lights, care must be taken that the pressure is not excessive, as, in addition to the result being extremely harsh, the fine tints will become thickened—battered, in fact.

In this case the cut is ruined. Again, over-anxiety to give the solids their full value must not tempt the operator to indiscriminately overload those parts with pressure, as in this case, also, the electro will be forced down by the heavy impression, and the very object of the machine-minder be defeated, as the surface of the plate will be pressed so low that the inkers will be unable to deposit the necessary quantity of ink.

As little work should be put into an overlay as possible. If it is overdone in the first instance, successive pieces have to be put up, and the probability of a successful print is very remote. As before stated, if the letterpress impression is satisfactory, and it is found desirable to patch the overlay, the pressure of the pages in close proximity is likely to be borne off, necessitating the placing of pieces on those parts affected. So it will be readily understood that this process may go on indefinitely, and at the same time yield a very unsatisfactory result in the end. It is not a question of the amount of actual work in the preparation of overlays, but of how the work is done. A man with fairly artistic taste will prepare an overlay which will " tell," in one-third of the time that another will occupy. We are free to admit that the latter may work very much harder and more anxiously ; but the value of efforts is judged by final results, not by the amount of labour those results have entailed.

FIG. 98.—FIRST PULL—FOUNDATION FOR OVERLAY.

FIG. 99.—FIRST CUTTING

FIG. 100.—SECOND CUTTING.

FIG. 101.—THIRD CUTTING

FIG. 102.—EFFECT OF COMPLETE OVERLAY.

FIG. 103.—CUT WITHOUT OVERLAY

FIG. 104.—CUT WITH OVERLAY.

FIG. 105.—CUT WITH OVERLAY PITCHED OFF.

CHAPTER XXII.

ROLLERS—THEIR TREATMENT AND MANUFACTURE.

BE the making-ready ever so well prepared, it is an admitted fact that, if the rollers to be employed in the printing are either badly made or out of condition, good printing is simply impossible. It has been stated, and with a great degree of truth, that the secret of good printing is a limited supply of ink and plenty of rolling —supposing, of course, that the overlaying, &c., has been properly attended to. Generally speaking, defective inking is answerable for more bad work than careless making-ready—a fact that should be constantly borne in mind by the machine-minder. Added to this, it is more easily recognised by the general reader. " Bad printing" is a vague term, but, in the majority of instances, when applied it is warranted ; and upon examination it will most probably be found that the censure is justified, not so much owing to defects in the printing proper, as from careless supervision of the inking—attributable to the bad or unfit condition of the rollers.

It may be said that roller composition is of two kinds, the ordinary material, made simply of glue and treacle, and the " patent," which is largely composed of glycerine in lieu of the treacle. Many printers prefer the original composition, while others are equally prejudiced in favour of the latest introduction. The "patent" will retain its virtue for a considerably longer period than the ordinary material.

Inventive geniuses have introduced from time to time

new compositions, each professedly being more durable and efficient than the two kinds generally adopted. India-rubber rollers have been often suggested and tried, and although it must be admitted that they were less liable to perish, they lacked that freshness and "tack" possessed by those mostly used. Composition rollers made with a large percentage of indiarubber are used in some newspaper offices, as the surface, while preserving its elasticity, is sufficiently strong to resist the destructive action of the rapidly revolving plates. But they cannot be said to answer the requirements of the book printer as, besides the disadvantages mentioned above, they are also apt to deposit almost all the ink on the first re-volution, which would render the colour of the work uneven. This defect is of course not accounted so serious in newspaper work—the primary anxiety in this case being speed.

A roller, to perform its work effectively, should be heavy, that the ink may be deposited thoroughly on the forme. Some stocks are made of wood, but iron is pre-ferable—not only because it is heavier, but because it will last very much longer. The wooden stock splits and perishes in time. The use of iron also often renders the riders unnecessary. The latter are, however, at times desirable, especially when heavy cuts are being printed, although they are apt to become destructive to the face of the roller after some little time, owing to the friction. In litho machinery the rollers are kept well on the face of the stone by springs; but with letterpress they must not be held down too rigidly, or the surface, coming into contact with the sharp edge of the type or plates, would soon be ruined.

It is never desirable to put a roller in work imme-diately it arrives from the maker's. In justice, however, to the manufacturers, we may say that rarely is the composition absolutely unfit; but it will be found that they will last longer in work, and prove altogether more satisfactory, if they are carefully stored for a few days. Roller cupboards are usually fitted along the walls of the machine room. A series of ventilation holes should be made at top and bottom of the sliding doors, otherwise the composition is apt to become damp. Iron or wooden racks are sometimes fixed in the open room for storage, but this plan is not so safe, as the faces are apt to become damaged. Never allow the surface of a fresh roller to rest against anything, or the pressure will cause a "flat," and render it absolutely useless.

A roller when in proper condition should be elastic without being positively soft. Prior to sponging up for use, press the fingers into the composition: if the indentation is marked, it may be placed aside as being too new.

New rollers are generally greasy, and lack that tackiness, "lug," or sharpness, which is necessary to insure their taking and depositing the ink satisfactorily. This quality is subsequently imparted by sponging with ley or water. In connection, however, with washing a roller, it should be clearly understood that it is necessary to wash *the surface* only, and not the body of the composition, which should by no means be allowed to become thoroughly saturated. The face is comparatively hard, owing to its exposure; but immediately under the skin the composition is absorbent, and if the damping or washing be injudiciously done, the whole is practically ruined. A roller should be *wiped*, not soaked, and the wiping done rapidly with a piece of rag. The top of the stock should be held in one hand, the other end being allowed to rest upon the floor, and the rag moved rapidly from top to bottom   If the roller has been in work, and ink has to be removed, strong *clean* ley should be first used, and afterwards clean water.   Warm ley is frequently applied ; this, however, is not advisable, excepting perhaps when strong-drying ink has to be removed, as the face becomes unduly softened, and the roller will ultimately " fret," or split, and small pieces of composition will be deposited upon the forme.   When a roller shows this disposition it should be at once discarded, or it will cause no end of trouble and bad work.   The defect will grow as the working proceeds.

The iron roller-trough is, in lieu of the washing machine, the best for cleansing purposes.   The extreme ends are rested on either side of the tank, and the sponging-up may be rapidly, easily, and effectively done. Machine-minders frequently sponge-up their own inkers prior to starting.   This is desirable, as they know, or should know, the proper amount of moisture, and the way to apply it, to insure a good tacky surface.   The finger should be passed along the face of the roller to ascertain its fitness.   It must feel rough, and offer some resistance to the touch.   If it be smooth, it is of little service, as it will take up a quantity of ink and deposit nearly the whole on the forme upon its first revolution.   If "tacky," it will take up the regulated quantity, and, while still retaining a fair supply, distribute it evenly over the surface of the entire forme, at the same time gathering up bits of fluff and extraneous matter, which, in the case of a " flat " roller, would be released upon the forme.

Nothing renders an inker so soon unfit as an excess of driers in the ink.   It speedily hardens upon the face, and forms almost an enamel covering, rendering the roller, of course, unclean.   When quick-drying ink is employed, the patent composition should be adopted. Turps, in lieu of water, is used for removing the

X

ink, and consequently the composition does not become so soddened. Added to this, the patent is much tougher and closer in substance than the ordinary glue-and-treacle composition.

A duplicate set of inkers should always be in readiness, especially in cases where the number is long. It will be found economical to give the old ones a rest at times, sponging and putting them in a cupboard to improve. By judiciously interchanging in this manner, rollers will last for some considerable period, and perform their work satisfactorily, whereas if compelled to run until they are completely unfit, their service is of short duration.

Rollers are apt to be considerably affected by the temperature, and also by the speed at which they are compelled to travel. In summer-time they are liable to become soft, and fret, and sometimes leave the stocks altogether. It is a much easier task to keep the rollers in condition in the cold weather than in the hot, as, if there is a tendency in the former case to harden, they may be sponged with lukewarm water, while the hardening process in the latter is altogether a different matter. When the temperature is high, constant attention is necessary, as if a small portion of a roller bursts, the whole is liable to leave the stock in a very short time, and a terrible mess and a protracted delay ensue. The patent composition is certainly to be preferred under these conditions, as it very rarely comes to pieces, being tougher and more stubborn.

Wavers and Vibrators as a rule may be worked for a longer period than the inkers, as their duties are not of such an important character. Not that we for a moment advocate looseness in their treatment, as proper distribution is essential to good inking. They should be slightly harder than the inker. The ends must be watched, as they are liable to split and fret. When they become at all hard or leathery, and relapse after having been wiped down, they must be discarded, otherwise the ridge of ink deposited by the vibrator will be simply reproduced at intervals over the table, and not properly distributed. The life of a vibrator naturally depends upon whether it has been cut to suit a peculiar forme. If portions of the composition have been removed to limit the supply of ink to various parts of the forme, it will be of very little service for use in the following sheet, as although the work may be the same, the cuts will probably be of a different character, or be placed in another line of pages, rendering the trimming for the previous forme altogether useless. From what has been previously said, it will be gathered that we do not favour the cutting of wavers to the extent to which it is frequently practised. The ductor screws are numerous, and their action is as effective as can be desired; and if the knife is true, and consequently presses against the roller, flush, by the additional aid of the leaden stops, the cutting of the vibrator should be of necessity an exceptional matter. If the portions immediately in a line with the gutters of a forme are cut away, of course the vibrator will be of equal service to the successive formes of the same work. But we question the desirability of cutting even to this extent, inasmuch as the wavers do not travel absolutely straight, but cross the tables at different angles. If a tightly fitting stop is placed in the correct position, very little ink really escapes. A very small quantity will most certainly "wander" to the space blocked out on the cylinder duct, but this is of little consequence.

With reference to the patent composition, as before stated, opinions are very much at variance as to its intrinsic qualities as compared with the original glue-and-treacle. The patent rollers are certainly to be highly commended for their general durability, as with ordinary care they may be used for months, and still remain in workable condition. They are usually *worn* out, which can be rarely said of the original—the latter generally losing its virtue and becoming "perished" in a short time. Prejudice, it must be admitted, enters largely into the matter, as when a printer has for years worked a peculiar kind of roller, and knows exactly how to manage it successfully, he is frequently averse to admit the claims of another material which requires different treatment. In large towns, contracts are usually entered into between the manufacturer of the ordinary composition roller and the printer to provide a machine with as many as are required for a term. Thus the printer may always discard a roller if it be in any way troublesome, and substitute a new one without incurring extra expense. With the patent, however, it is somewhat different, as the cost is dependent entirely upon the number of renewals. This necessitates the machine-minder taking every possible precaution to keep the roller in proper trim ; and if ordinary care is used we have no hesitation in saying that they may be relied upon for all kinds of work for several months. A single inker of the ordinary composition is sometimes introduced at the leaving end, as it will take up small foreign particles from the formes, which the patent material often fails to do.

Turps must be used in lieu of water or ley for cleaning the patent rollers, as the latter would speedily destroy the surface. The cost of turps is of course very much greater than that of ley, but it must be remembered that it is not absolutely necessary to wash these rollers when the stoppages are short (at meal-times, for instance). But they should be lodged in a closed cupboard to prevent the accumulation of dust and grit upon the surface.

Excepting in the case of a tolerably extensive printing office, wherein a large number of machines are employed, we very much question the economy of making rollers. Taking into consideration the necessary sacrifice of space, the plant required, and the wages to an experienced maker, we feel assured that the saving is very problematical. Several other matters have also to be considered. Inferior rollers—rollers that would be unhesitatingly returned to the ordinary manufacturer—are often supplied, and the fact of their having been made on the premises necessitates their being used, or the machine-minder may lay himself open to the charge of being prejudiced. Again, even in the event of their being in every way satisfactory, there is a greater inclination to use them more liberally, as the fact of their being made by the firm fosters the idea that new ones cost nothing. The above circumstances, taken in conjunction with the constant cost of new material, are apt to render roller-making an expensive matter to the printer.

These remarks, of course, apply more particularly to large towns where roller-making is carried on as a business. In isolated instances, however, it is frequently absolutely necessary that the printer should manufacture his own rollers, and for the benefit of such we append the following hints.

It must be remembered that the materials used should be thoroughly good, as otherwise the composition of necessity will be inferior. The glue should be clean in appearance, tough, and brittle.

In the first place, the glue must be placed in a vessel, and sufficient water poured in to cover it. Allow it to remain till about half soaked through, and pour away the water, allowing the glue to remain until it is perfectly soft. In this state it is ready for the melting-kettle. This kettle somewhat resembles the ordinary glue-pot, *i.c.,* it consists of two vessels, the inner of which contains the glue, and the outer, water.

The boiling may be done by means of gas-jets arranged beneath an iron slab, the kettle being dropped into a space specially made for its reception. The water in the outer vessel should not be allowed to boil, or the composition will be injured.

When the glue has been thoroughly melted, the treacle may be mixed in, and the whole boiled for about an hour. Before putting in the treacle, be sure that the glue is not too thick. If this should be the case, add a little water and well stir. The mixture should be allowed to boil for an hour, but care must be taken that the boiling be not overdone, as the treacle is apt to harden, and naturally will affect the composition, causing it to lose that "tack" which is so essential to a good roller.

In mixing the composition, it will be found advisable to slightly alter the proportions, according to the time of year, or the peculiar class of work the rollers are required for. In cold weather it is advisable to add a larger proportion of treacle, as this ingredient tends to render the composition softer. Press rollers may generally contain more treacle than those required for quick machines, as the former are more tenderly used; besides which a roller is really better for a slight excess of treacle, while too much glue, although imparting solidity and strength, is apt to cause it to become hard and stubborn after little work. The composition should not be stirred excessively, or bubbles or froth will be produced, which may be eventually found in the newly cast rollers.

The fitness of the composition is usually tested by placing a small portion on a piece of paper and allowing it to dry, when it should feel firm to the touch. If an indentation is not left by the finger the composition may be regarded as ready for pouring.

The core or stock of a roller should be carefully cleaned before placing in the mould. The old stuff must be, stripped, care being taken that the smallest pieces be removed. If the core is made of wood, it should not be wetted, as the presence of moisture prevents the composition from adhering. It is advised, when the composition shows a tendency to leave the stock, to wash the latter with a solution of quicklime and water, allowing it to dry, however, before pouring.

The stocks are frequently painted with ordinary lead paint, which helps to preserve them.

Before removing the old composition, the face of the roller should be well washed with hot water, and the surface removed with a sharp knife.

Before pouring, the mould should be slightly warmed and well oiled, to admit of the roller being easily removed when dry. At the same time it should be remembered that an excess of oil is apt to cause an uneven surface. The bottom piece should also be oiled, and the stock placed in position. The composition must be poured slowly on one side of the mould, to allow the air to escape on the opposite side.

After filling the mould, allow it to stand in a cool place for at least twelve hours, when it should be in a fit and proper condition to be drawn.

The labour involved in drawing will greatly depend upon the manner in which the cleaning and oiling of the mould have been attended to. Frequently, however, the great difficulty experienced in this operation is owing to the make of the mould itself. The tube is sometimes slightly bent, or the planing of the inside is faulty, and in each case some amount of time is expended in releasing the roller. In this business, as in most others, special methods are adopted by different makers. One firm have their moulds constructed upon an entirely new plan, and after the casting has lain sufficiently long, the mould is hung up in such a manner as to allow the roller to release itself.

The composition of a newly made roller covers the extremities of the stocks. The ends of the latter may be easily ascertained by feeling, and the superfluous stuff cut away by means of a thin wire or piece of string. If the composition shows a tendency to be loose on the stock, dip the ends in hot water.

If, when the roller is taken from the mould, the surface is uneven or ragged, it may be improved by washing with lukewarm water. But if it is really defective, strip the composition and recast.

Before re-melting old composition, cut it into small pieces and place it in cold water for about an hour. A sieve is generally used for this purpose. Composition will become poor, or perished, by continual recasting, but the judicious addition of treacle and glue (a larger proportion of the former) will restore the bulk to a fit condition.

In recasting the "Durable" composition, the stuff should be mixed with its own only—that is, no single ingredient must be added, or it will be spoilt. The old composition should be well cleaned with hot ley, and cut into strips. Weigh the old material, and first put into the kettle new composition equal to a quarter to one-half the weight of the old. The exact proportion must be dependent upon the state, or the length of time the latter has been in use. When this is thoroughly melted, add the old and mix.

ROLLER COMPOSITION.
Hansard's.

Glue	...	...	...	...	2 lb.
Molasses ...	...	...	...	6 lb.
Paris White	...	...	...	½ lb.

Glue	...	...	...	...	12 lb.
Refined Sugar	...	...	...	10 lb.
Glycerine ...	...	...	...	12 lb.

The glue to be first melted, and the sugar and glycerine added.

Glue	...	...	9 parts by weight.
Treacle	...	...	14 ,, ,,
Paris White	..	1 ,, ,,

Glue	...	...	...	... 4 parts.
Golden Syrup	...	...	... 7 ,,

The majority of prepared-roller manufacturers now sell the composition, properly prepared, by weight. In this case the printer has merely to remelt and cast— the stuff being improved from time to time by the addition of virgin material.

The following are the names of London firms who make the manufacture a speciality :—

MESSRS. HARRILD AND SONS.
"DURABLE" COMPOSITION COMPANY.
MESSRS. DUDLEY.
MESSRS. FLEMING.

CHAPTER XXIII.

PAPER: ITS TREATMENT — STORAGE — WETTING —
ROLLING—ROLLING MACHINES.

PAPER always has been, and presumably always will be, one of the professed fruitful sources of failure in printing. In the event of things going generally wrong, the paper is very often blamed. Rightly or not, paper, in conjunction with rollers and ink, is generally made answerable for more disastrous results than are the making ready and general working of the machine. Without being in any way uncharitable to the machine-minder, and while fully admitting that the condition and quality of paper at times considerably increase the difficulties to be contended with, it must be confessed that it often receives more censure than is warranted.

Paper thoroughly good and fit in itself may be rendered almost unworkable by careless treatment. When it is worked dry, or in the same condition as delivered from the mill, it will as a rule be found to give little trouble. With the *quality*, *colour*, and *substance* we have comparatively little to do. These, of necessity, depend entirely upon the price paid, regulated by the requirements of the class of work to be done.

At the time of writing (1888), paper is astonishingly cheap, and inasmuch as really good material suitable for first-class book-work may be obtained for 3¼d. or 3½d. per lb., really low quality is used only for the commonest publications and newspapers. Printers are now provided with paper of good substance and colour for ordinary cut-work, whereas, some years ago, loose, open-grained, and badly surfaced material was frequently supplied, and satisfactory results expected, but very rarely, it must be admitted, attained.

Plate and heavy paper should be delivered "flat," every ream or half-ream being well secured between battens. On delivery into the paper-room or warehouse, it should be unpacked and carefully stacked, a wetting board slightly larger than the sheet being placed at intervals. The edges must be protected by sheets of wrapper. This is especially necessary at the base of a stack, as otherwise the edges are liable to be kicked, or damaged by passing trollies, &c.

It is of course necessary to use a ladder when stacking, after the paper reaches a certain height. Inasmuch as it is heavy, a special pair of steps about four feet high should be constructed for the purpose, being both firm and strong. The platform at the top must be sufficiently wide to admit of the porter standing with a feeling of safety. The back supports should be nearly perpendicular, to admit of it being placed almost flush with the stack, that the man may have more command over his work. Folding steps and single ladders should be avoided. The former are unsafe, and the resting of the latter against the heap is apt to injure the edges.

In speaking of the storage of paper, it may be mentioned that the room should be perfectly dry, otherwise the material will absorb the moisture, and become rotten and of bad colour. The basement is the best for the purpose, as the safe delivery is facilitated. The floor should be concreted, and along the walls may be erected a continuous stand, about twelve inches from the ground. This will prevent any damp reaching from the underneath, and the place may be easily cleaned.

Paper of large dimensions is frequently sent in loosely and doubled over in three. This is a good plan, facilitating the handling during its transit from the mills, and preventing the formation of a "back" or crease in the centre, which will occur if folded in quires. Upon delivery it should be released from the wrapper and laid out flat, a wetting board with a heavy weight being placed on the top.

New or "green" paper should be avoided. As a rule it is despatched from the maker's in this state, but it is advisable to let it remain to season for at least a fortnight or a month before printing. Sometimes it is not so thoroughly dried as it should be, and is liable to crease in the printing in consequence. Paper, when new, will sometimes be found to be dry in the centre of the sheet, and slightly spongy towards the sides. This defect is often caused by careless finishing at the mills. It perhaps may not be discovered if the forme be an open letterpress one, but if a large tint or border rules are being printed, an ugly crease will inevitably result, as there is a slight contraction in the centre of the sheet. In this case, register coloured work is altogether out of the question, and no amount of care bestowed on the forme will insure uniform fitting. By "tapping" a plate, perhaps a few sheets will be "in," but the work as a whole will be a failure.

Makers are not always to be blamed for the unfit condition of paper when first delivered. It frequently happens that a peculiar job has to be produced in a hurry. The paper of a given size and weight must be made, and, under pains and penalties, is delivered in an almost impossible time. The printers' difficulties now commence, varying of course with the style of work. The paper has been unduly hurried in the manufacture, and is green and imperfectly dried. If heavy cuts are to be printed, it will possibly be discovered that the surface is not sufficiently hard and firm to hold the necessary quantity of ink, while the face becomes damaged. If an attempt is made to print in several colours, the paper will stretch unevenly, and no amount of rolling will thoroughly take out its elasticity. Paper to be in good working order should have sufficient time allowed for the making, and be stacked for some weeks prior to use, in order to allow it to mellow.

Ordinary printing paper is wetted or damped prior to printing. Highly glazed, or super-calendered, and plate paper, are worked dry. The latter would lose its bright, hard surface, and often, being extremely soft, would be entirely ruined by the application of water. We will first deal with the former. By the greater or less quantity of size in the composition, the maker can, to a nicety, regulate the hardness of the paper. As a rule, the better the surface the harder the material, as, in the making, greater pressure is exerted in the finishing process. When it is desired that a paper shall " bulk " or " handle well," *i.e.*, feel thick, the material is subjected to less pressure in the first rolling process, and is consequently softer than that of a thinner or more highly finished sheet. It is obvious that in the damping this will absorb more moisture than the thinner paper, although the weight in each case may be the same.

The degree of softness may readily be determined by allowing the tongue to touch the surface of a sheet. With a hard or highly sized paper the moisture will remain on the face and show no inclination to become absorbed. With half-sized papers, there is a greater clinging to the tongue ; but with soft and plate papers the moisture is immediately taken up, and the surface slightly blisters, showing that damping would be out of the question, even supposing it were desirable.

THE WETTING TROUGH

ordinarily adopted consists of two separate parts—one being for the heap of dry paper, which is raised higher than the other containing the water. Both are lined with lead. They are of course made of various · dimensions, but it is advisable to have them of a full size, as it is necessary that the wetter should have plenty of space when dipping large sheets. It is desirable to have a tap fixed immediately above the water trough, and it is also a good plan to provide a pipe at the bottom, with a plug, that the water and any sediment may be thoroughly and easily emptied when required.

In dealing with ordinary printing paper, place a wetting board at one extremity of the water tank, upon which a piece of wrapper may be spread, together with, say, five sheets of the paper. Several reams to be wetted down having been laid (the narrow end to the right and left) on the higher portion of the trough, take, say, eight sheets, holding them firmly at either extremity by both hands, and pass them under the surface of the water. It is needless to say that, as in all matters appertaining to printing, to do this successfully and rapidly, a certain amount of experience is necessary—the chief difficulty in the dipping proper being, perhaps, to hold the whole well together without the edge of the top and bottom sheet turning over. By raising the front part of the paper immediately prior to lifting it from over the water trough, the water will run off at the leaving end. The whole should then be laid *squarely*, upon the flat sheets already in position, and a second lot be taken from the dry heap and treated in the same way. It is impossible to say, with any degree of accuracy, how many sheets should be taken for a single dip, as this entirely depends upon the quality and relative hardness of the paper, and also upon the character of the work to be printed. If the paper is well sized, it may be necessary to dip in sections of six or seven sheets, but if softer in texture it may be taken in half-quires. With fairly soft paper, two heaps of dry material may be used to facilitate the wetting— the one in the ordinary place, and the other by the side of the wet paper. When this plan is adopted, a section, after being well damped, is placed, and another from the second dry heap put immediately on the top—to be followed, of course, with another wetted section. As we have before said, paper is very absorbent, and readily imparts its moisture, which process is facilitated by heavy pressure. When large quantities of paper have to be wetted, it is advisable to have substantial trollies made, with wide flanged wheels, upon which the newly wetted paper may be stacked. When a given quantity is finished, the whole may be wheeled into an out-of-the-way corner, and left to mature, an empty trolly being placed in the position of the previous one for the reception of the next lot of paper.

The extent to which various appliances may be adopted in the wetting must of course be regulated by the amount of work of this description to be done. Paper in bulk is extremely heavy, and inasmuch as in transit, especially after wetting, the utmost precaution should be taken to prevent injury, it is advisable to employ those means by which it can most readily and safely be conveyed from one position to another with the least possible handling. We may mention that at the Belle Sauvage Works, London, the system adopted is as perfect as in any printing office in the country. Practically, the paper may be said not to be handled in bulk after it is taken from the stacks into which it is made immediately upon delivery. A series of trollies are employed, having flanged wheels, and tracking the entire wetting rooms upon counter-sunk rails. At various points junctions are made, so that the truck may be drafted off to almost any desired position. On the top of the trolly-board are grooves, upon which runs a wetting board, fitted with flanged wheels. The supports of the stacks of paper are so constructed as to be on the same level as the travelling trolly, so that large quantities of paper may be moved from any position by simply rolling the whole from the support on to the trolly, and the latter be pushed easily to the desired position. By this contrivance stacks of paper of considerable weight may be deposited in position after wetting, with little labour and without injury.

As before stated, newly damped paper should be subjected to some considerable pressure. A board placed on top of the heap must be surmounted by heavy weights, and allowed to stand for, say, twelve hours. If at the end of this time it is still wavy, or has unequally absorbed the water, it should be " turned " and again left to mature.

If paper is found to be too wet or soddened after standing, in the turning fresh sections of dry paper should be introduced at intervals.

For the commonest descriptions of printing, such as newspapers, &c., it is desirable to have the paper thoroughly damp, as the process of making ready is simplified, the paper more readily adapting itself to the surface of the forme. Added to this, the soft material will "dip," and take impression from defective letters, &c., slightly lower than the main body of the page. Of course in this class of printing the ink used is very thin, as stiff ink would tear the surface of the paper, but being more elastic, the forme may be worked with a lighter impression.

It will be gathered from the above that for fine printing the paper must be relatively harder, to enable it to resist the "lug" of stiff ink. Otherwise the surface of plates would speedily become filled with "picks"— caused by pieces of paper being held by the plates.

It has been stated that surfaced and well-sized paper requires to be dipped in smaller sections than the softer qualities, because it is not so absorbent. Care must, however, be exercised that the surface is not injured, which it inevitably will be by allowing it to remain too long in the water.

When wetting, a sheet should be cornered every 250 sheets, *i.e.*, the corner turned down to overlap the edges. This is termed the token-sheet, and materially facilitates counting.

Machines are sometimes employed for damping, the most successful, perhaps, consisting of a large cylinder (covered with blanket) revolving in a trough which is filled with water to a regulated height. The sheet or sheets are fed between a series of tapes and carried over the large drum on to the other side, absorbing a given quantity of moisture in transit.

The majority of book-work printers, however, prefer to wet-down by hand, no matter how great the quantity. An expert will damp, say, ten reams of double-demy per hour, and he is able to give that special attention to peculiar paper which would be impossible in a machine.

Y

It may be mentioned that the addition of a small percentage of glycerine to the water will materially assist the drying of the ink after printing.

Reels of paper for rotary machines are of necessity damped by a special apparatus. This has been described in connection with newspaper machines.

Presses are frequently employed for purposes of pressing newly damped paper. The ordinary screw-press is well adapted for this purpose. Care must, however, be exercised in the regulation of the pressure, as, if excessive, it is apt to drive the water out at the sides, while that in the centre being confined renders the sheet of unequal dampness, the edges being the driest. Experience will teach the proper " nip " to be applied. We do not advocate the use of the press generally, as with proper treatment we think paper becomes more fit if left in the heap after wetting, with simply the ordinary boards on the top and a good weight, than by being subjected to an unnecessary pressure. If only a short time be allowed, then the press may be profitably employed, but the paper will not be in such a good condition as if allowed to stand in the ordinary way.

Paper for fine work is frequently damped and rolled between zinc plates prior to printing, to insure a super-fine surface. Of late, however, owing to the additions to the paper-making machines, the paper is delivered with a splendid surface, which renders this extra expense unnecessary. But it is desirable to have a rolling machine on the premises, as paper is at times materially improved by being passed between zinc plates. This remark especially refers to paper required for colour printing, when it is necessary that every particle of stretch be taken out. It should be rolled twice each way, backwards and forwards through the rolls—the narrow way of the sheet, then turned and allowed to run broadside.

Speaking of colour-work, we may say that "green" or new paper should never be used. If it is not allowed to stand and mature for several weeks, it is probable that no amount of rolling will thoroughly take out the

Fig. 106.--Hand Rolling Machine.

stretch. Be the paper ever so good, or even guaranteed by the maker, if it is put into work immediately after it is made, innumerable difficulties will be sure to crop up in the working. The best material may have been employed, but paper newly made is naturally spongy, and should be allowed to harden. If not, it will "go" in various directions—some sheets being wide, others long—and no

Y 2

alteration of forme will render the work satisfactory. When many colours have to be printed, it is advisable to roll even enamel papers.

By means of the powerful screws it is possible to exert almost any amount of pressure when rolling. The aim should be to put on a surface (for ordinary work) without materially reducing the substance of the paper. Paper may easily be *crushed*, but this is not rolling. Greater pressure is of course necessary for material required for register work. The appearance of paper of low quality is not much improved by rolling, as pressure brings to the surface all the imperfections in the shape of various coloured specks, &c. ; and although a good face may have been imparted, the sheets will be dirty and spotty.

Rolling machines consist of two bright chilled steel rolls, fitted, one above the other, in a powerful side-frame. An iron table is placed immediately in front, the surface being level with the top surface of the under-neath roller, upon which are laid the zinc plates having the paper between. On the other side of the roller is another table, which in some instances has an upward tendency, so that when the plates are run through they may return between the rollers by their own weight, obviating the necessity of having a boy to push them in on the reversal of the machine. Where a large quantity of rolling is done, both tables are made perfectly flat. Two sets of boys are employed—one at either end. While one set is filling in, the rolls are utilised by the other, who have the plates ready. Immediately in front of the machine should be fitted a substantial counter, sufficiently large to take the largest sheet lengthwise, and three times the width, allowing for space for side-marks against which the plates or paper may be laid. On one side we have the stack of white paper to be rolled, in the centre the plates, and on the other extremity the sheets that have been removed from the plates after rolling.

The modern rolling machine is far less cumbersome

Fig. 107.—Rolling Machine with clutch Motion.

in construction than those which were in use a few years ago. It is better geared, running with comparatively little noise; and the clumsy clutch motion is superseded by the substitution of one fixed rigger, on either side of which is a live pulley. One strap is crossed, while the other runs straight. By moving the striker, either strap may be led on to the fixed pulley, the fork at the same time leading the other band on to its loose rigger. As has been stated in a previous chapter, the strap should be four-inch double, and have a fairly long travel, which will to a great extent obviate slipping.

Only the man in charge should be allowed to place the plates in the machine and reverse the rolls. These are very dangerous machines : mishaps soon happen, and with the rolling machine they are usually of a very severe character.

A distinct set of plates should be kept for every size of paper required to be rolled, as they become indented in time at the position occupied by the edge of the sheet, and are liable to mark sheets of a larger size. When ordering plates it should be understood that they are at least four inches larger both in breadth and length than the size of paper for which they are made. Thus a double-crown plate should measure 34 in. by 24 in. There can be no doubt as to the economy of procuring the best quality of zinc. Inferior makes are liable after a little wear to flake, and the pieces become indented in the rolled paper, causing serious batters in the printing. The edges should be slightly rounded with a pair of strong shears, as the sharp corners' render them dangerous to handle, especially when new. When they begin to wear they must be discarded, the stoppages necessitated by pieces flaking, &c., causing considerable delay and defective work. The price realised by the sale of the old metal largely contributes to the cost of a new set.

Zinc plates should be frequently cleaned with rag, as dirt and grit are apt to accumulate on the surface. The ceiling above the machine and feeding-in counter should be match-boarded and swept periodically, as troubles often occur from flakes dropping upon the plates or rolls.

Before rolling printed work, be sure that the ink is perfectly dry, or a large proportion will come off on to

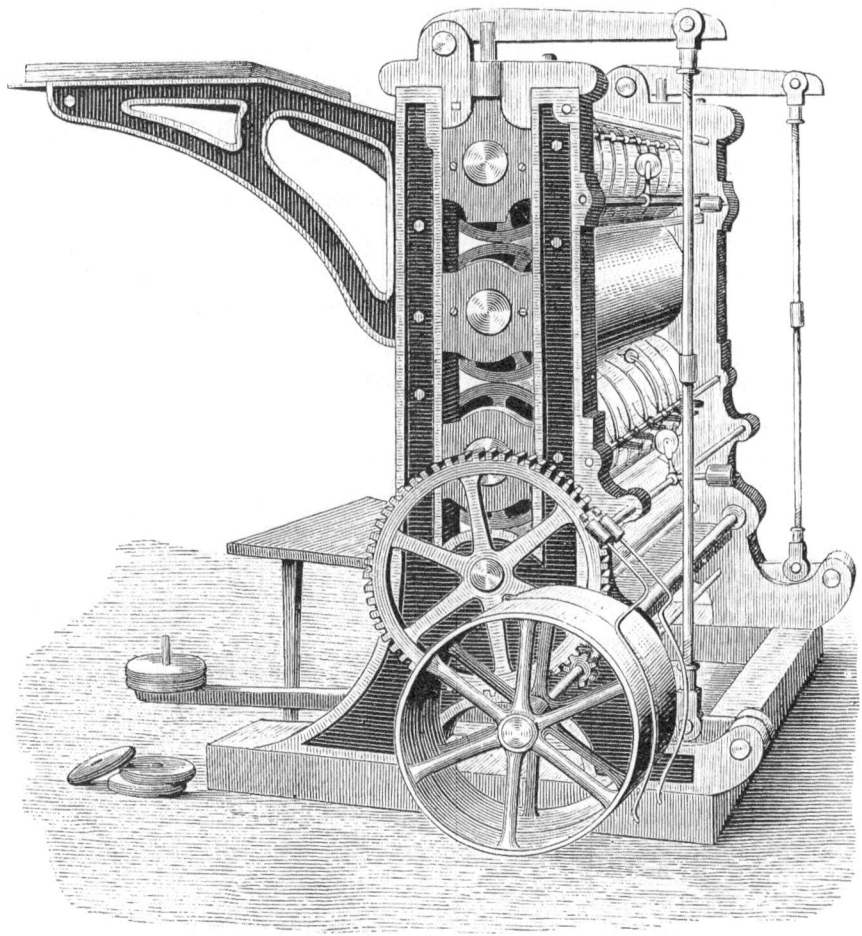

Fig. 108.—Heim's Rolling Machine.

the plates, and become deposited upon the following sheet. Even when the ink is hard and set, it will be found advisable to frequently rub the zincs, as a dirty film will form on the surface.

For smoothing and putting a fair surface on paper

the rotary machine manufactured by Heim and Co., of Offenbach, is excellently adapted. The machine consists of four geared cylinders fixed horizontally above one another in a side-frame. The top and bottom rolls are of polished steel, while the two centre ones are surfaced with compressed paper. The paper is laid on a board above the top cylinder, and the sheets are fed-in singly, travelling over the first steel roll, receiving a nip between this and the paper cylinder immediately underneath. Passing between the two centre (paper-surfaced) cylinders, the other side is rolled immediately before running on to the taking-off board between the steel and the roll immediately above. The impression is regulated by set-screws and balance-weights.

The paper should be slightly damped before rolling by this machine. It is unsuitable for hard and "tinny" material, which is apt to crease. If the object of rolling be rather to take out the stretch than to put on a surface, the ordinary machines are to be preferred.

The power necessary for driving is comparatively slight. Two boys are required—a layer-on and a taker-off. Of medium-sized paper—say double royal—about 1,000 sheets per hour may be rolled. These machines are excellent examples of engineering work, the fitting of the gearing, &c., being perfect. Although four cylinders are working in gear, they are practically noiseless.

CHAPTER XXIV.

COLOUR PRINTING.

By Letterpress Process—Various Methods of Production from Plates—
Registering—Wood Blocks—Process Blocks, &c.—Best Machines
for the Purpose — Inks and Ink Making — Bronzing and
Bronzing Machines.

A COLOURED design of any description may be reproduced by the lithographic or by letterpress process. Both
methods are, of course, largely adopted, and it is a matter
of dispute at times as to which is the better and the more
economical to select. Up to a very recent period nearly
the whole of the presentation plates, frontispieces, &c.,
were printed from raised blocks, but gradually litho-
graphy asserted its superiority in many respects, and
we think we are correct in saying that the latter process
is now adopted in nine cases out of ten.

In comparing the rival methods, very many points
have to be considered. Given a limited number of
printings, it will generally be admitted that a better
result can be obtained from the stone. The reproducer
is enabled, by means of fine chalk-work or stippling, to
give greater value to a colour than by an etched plate
or wood-block. His work is less liable to injury from
the printer, as with ordinary care the finest artistic
touches are preserved, whereas in the case of the raised
block, the lines, &c., are apt to become injured by a
too heavy impression, or to disappear altogether from injudicious cutting away.

Again, there is the original cost of the drawing to be
considered. When a line is made by the artist upon

the stone, a fac-simile can be produced on paper by the printer. But in the case of block printing, this line has to be either made on a wood-block and engraved, or transferred to zinc and bitten up, before it can be printed from. Duplicates for letterpress printing are electrotyped from the original block, a very much more costly and uncertain process than the transferring upon stone in the litho. We say uncertain, as the plates are frequently distorted and irregular, owing sometimes to the careless moulding. Stereotyping in connection with colour blocks is, of course, entirely out of the question, as the shrinking of the metal is both great and irregular, owing to the contraction of the matrix when subjected to the heat of the molten metal. Moreover, it is impossible to obtain the necessary sharpness by means of a *papier-mâché* mould.

Printing from block is undoubtedly a quicker process than from stone, although the making ready takes longer, especially in the case of formes made up of isolated pieces.

Again, success in the registering is very much more dependent upon the boys in block than in litho printing. Given the paper in first-class condition, the laying-on is a much simpler process in the latter case than in the former. The sheet is laid upon a board at an angle, and the front edge rests flush on to the gripper-bar by its own weight, while the side-mark moves slightly forward immediately before the sheet is secured, to insure the side lay being perfect. With the letterpress, however, the sheet is lifted from the heap on the bank, taken in the centre at either side between the thumb and first finger, and placed on the points, care having to be exercised that the front edge is perfectly flat in the cylinder entry. If the boy is not well trained and quick in finding the points, he has either to strike off the impression or to allow the sheet to be spoiled.

Another fruitful source of spoilage is the tearing of the point holes, mostly caused by careless laying-on. The sheet has to be simply *laid* on the spurs; if force is used, the hole will be fractured, and defective fitting must necessarily result.

It may be mentioned, however, that in block printing there is less probability of the paper becoming troublesome through stretching, because there is much less pressure, owing to the surface being raised, and also because the impression is given only on the relatively detached portions of plates. In the case of litho an even pressure is necessarily exerted over the entire sheet, which tends to stretch the paper in the event of its not having been suitably rolled.

It must not be supposed that letterpress colour printing is free from troubles with reference to bad preparation of paper or from material wholly unsuitable to the purpose. However carefully a forme may have been registered in making ready, it will sometimes be found that the sheets will vary considerably when working the bulk. Atmospheric changes will materially affect paper, especially if it has been allowed to lie in small lots, offering greater surface to the changing temperature. With superfine, well-rolled, or specially prepared paper, however, little difficulty should be experienced in this respect. It is only the soft, spongy material that causes so much trouble—paper which absorbs the damp and stretches in wet weather, and, as a natural concomitant, becomes proportionately dry and shrinking in the warm. If common paper is of necessity used, it should be allowed to lie for some considerable time after delivery, and ultimately be rolled.

Makers should invariably be taken into confidence when paper is required for register work. Material that

will answer every purpose for black work may be found
to be thoroughly unsuited to receive a succession of
tints.   The surface may be too hard, and in this case
the ink will stand up and present a glossy appearance,
and the following colours will not pick up.   It will be
found that the deposit will be scabby or uneven, owing
to the face of the paper being practically enamelled.   It
must not be supposed, however, that the paper should be
absolutely soft.   Half-sized material is the best adapted
for this purpose, and the colour will then be slightly
absorbed without being dull.   More ink is required, but
the cost of an extra percentage of colour is a far less
serious consideration than are perpetual stoppages and
unsatisfactory results.

With reference to the production of colour blocks,
various methods are used, and it may be cited as a
singular fact that almost every printer is convinced that
his peculiar plan is superior to those adopted in other
establishments.

The subject is sometimes entirely produced on engraved box-wood blocks.   This plan, however, is expensive, and only admissible when the design is comparatively
small and the number to be printed is large.   The " key,"
or outline, is drawn upon wood, and when engraved is
very similar to an ordinary woodcut, excepting that extreme solids are more sparingly used and the tinting is
much more open.   Just sufficient work is put in to give
shape and life to the design, allowing the successive tints
to supply the necessary depth.   The artist, however, often
puts plenty of work into his drawing, as by so doing he
materially decreases his labour in the following colours,
which oftentimes may be entirely flat, inasmuch as the
key " cuts up," or gives the necessary shape and brilliancy
to the whole.   An experienced artist rarely leaves much

"drawing" proper to the colour blocks, because if he finds in the proving that the subject is coming too dirty or pronounced, the colour of the key may be materially lightened, and thus the defect may be obviated. If too much, however, is left to the successive blocks, the result will probably be unsatisfactory. In some cases, where the number of printings is a matter of no importance, both a key and "touch" are employed, the latter being simply for the purpose of giving depth to various bits in the design, thus admitting of the drawing block being printed in a light colour, which imparts, of course, a large degree of softness to the general appearance.

Although good effects may be produced by a series of engraved wood-blocks, unless great care is exercised in the cutting of the tints, or an unlimited number of printings is allowed, the finished print is apt to appear hard or stiff. Very frequently a wood key is adopted, the supplementary colour blocks being produced by the aquatint process. The tints, again, may be drawn upon stone. Transfers may be taken from the chalk drawings upon zinc plates and bitten up. Although it must be admitted that of late the surface plates so produced are in every respect superior to those supplied a few years ago, unless the chalking upon the stone is both firm and open, a large amount of the work is found to be wanting, when an impression is pulled from the zinc and compared with a print taken from the stone. This is not to be wondered at, when we consider that some lithographic artists are prone to make their work as filmy and delicate as possible, and throw the responsibility for the reproduction upon the workman whose duty it is to transfer it to the plate and bite it up, at the same time allowing sufficient depth to insure good printing. It is a well-known fact that transfers taken from a chalk drawing, and put upon another stone to be printed from by the lithographic process, are often very defective, the lightest parts often disappearing, while the more solid portions are apt to become thick and muddy. It will therefore be understood that, when the biting process follows, these defects are materially intensified, and, while much of the finer portions of the work is lost, very frequently hard and harsh lines appear around the extremities of the drawing. Every plate should be examined carefully, and judiciously touched by the engraver before it is proved.

When the colour blocks are first drawn upon the stone, prior to biting-up, although it is advisable, not to say necessary, that each should be rolled-up in colour by the litho printer, and proved in the ordinary way, the success should be judged from the proof from the *blocks*, not from the *stone*. The artist may refer to the litho impressions, and bemoan the great difference that exists between the letterpress and the litho proofs, but this more intimately concerns the artist himself than the printer, as the latter is simply responsible for producing fac-similes of the print from the blocks. A printer, when receiving a proof with the intimation that the bulk is required to be equal in quality, should be perfectly sure that the print is thoroughly legitimate—*i.e.*, printed from the blocks that will be supplied to him. We mention this because it is the custom of some houses to employ artists to reproduce a chromo, without stipulating that the proof submitted shall have been pulled from the surface blocks, the result oftentimes being that the print is practically a *litho*, having been pulled from the stone. When the plates are supplied they are frequently woefully inferior to the litho impressions, and trouble follows.

As we have before mentioned, much better results

may be obtained from stone than from blocks in the same number of printings. The designs of some artists, however, are peculiarly suited to reproduction by the block process. When a strong outline or key is made, and the colours or tints are comparatively flat (taking Miss Greenaway and the late Randolph Caldecott as examples), the difficulty in producing a fac-simile in a limited number of printings is comparatively an easy matter, simply because there is but little shading or detail, excepting in the drawing or key block. In this peculiar but popular style of chromo work, the key block may be either drawn and engraved, or put upon stone and bitten up, the effect in the latter case being in every respect equal to the engraved block, as the work is very open. With ordinary care, the zinc outline plate should in every particular be perfect, as the process is singularly adapted to open-line work. It is when fine tinting is adopted that the requisite sharpness is found to be wanting. Of course it will be readily appreciated that the difference in the cost of the bitten and engraved block is very marked, the former being considerably less expensive.

With small subjects, of which long numbers are required, the system of drawing upon box-wood and engraving in the ordinary way is frequently used. As in all work where only a few printings are allowed, a good full key is absolutely necessary. Care must be taken by the engraver in the engraving that the tint line should run in the direction which will produce the best effect, and the artist should indicate the same. If this point be studied, a pretty texture may be imparted; if, on the other hand, it be neglected, the result will be very poor, very considerably lessening the intrinsic value of the work.

In chromo work it should be understood that extreme

softness can only be obtained by a number of printings. By this we mean that to insure general softness, it is necessary to employ a succession of delicate tints. If, for instance, it is stipulated that a fully coloured design be reproduced in, say, eight printings, the probability will be that an attempt will be made on the part of the artist to insert an excess of work into every block. The result will be that we shall probably have either a harsh and highly coloured proof, or one that is poor and weak, lacking character and effect. Not that it is desirable to give *carte blanche* to the artist, in which case the number of separate blocks required might possibly amount to, say, fifteen or sixteen, or even more, and the ultimate cost be excessive. Some reproducers are competent to reproduce excellent effects in fewer colours than others, owing to their being *artists* proper rather than mechanical artists, who are afraid to give sufficient value to every block. However, this branch of colour-printing perhaps rather concerns the individuals whose duty it is to arrange for the original blocks than the mechanical colour-printer. It must be remembered that the printer is only able to produce effects strictly in accordance with the materials afforded. To the artist belongs the duty of supplying the finished proof to be imitated in bulk.

Engraved blocks cause less trouble to the printer than those produced on zinc, as the latter are usually shallow, and lack that sharpness characteristic of the former process, in addition to which, as we have before mentioned, the extreme edges of the tints are very often hard, and require the utmost care in the making ready. Electros from wood-blocks are also much more satisfactory, owing to the sharpness and depth.

When proving the original blocks at press, a "book,"

or series of proofs of the individual colours, is pulled together with the combination of the whole up to the last colour. When the subject is not too large, it will be found convenient to have the entire set stitched together in a strong wrapper, arranged 1, 2, and 1 and 2, 3 and 1, 2, and 3, and so on. By adopting this plan references are always to hand, and there is less probability of their being mislaid.

Although the key or outline block is not always printed first, it is necessary to have this primarily and correctly imposed, in order that the various tints may be properly registered. Obviously it would be impossible to lay down the first set of plates, say buff or yellow, without any guide or proper indication for their exact position. Therefore, the key, which may be in brown, grey, or black, is first properly imposed, with due regard to correct furniture, lay, &c., and an impression pulled upon transparent paper, that the tint plates may be laid upon the blocks and secured exactly in their proper position. Mahogany is perhaps the best material for mounting these plates upon, as it is hard and not liable to warp. In laying down the tints, the pages or various portions are placed on the block, and by putting the oiled sheet on the surface they may be shifted with tolerable accuracy to their correct position. It is a far more satisfactory and a quicker method to impose the plates upon the stone or substantial table than on the bed of the machine. To insure perfect register a great deal of care is necessary, and it is therefore advisable to have the entire mounting block under convenient control, which can hardly be said to be the case when it is fixed into the coffin of the machine. Although it is very possible that several alterations will be required to be made when finally making ready after underlaying, the machine-minder

z

should endeavour to secure the plates as accurately as he can, or the tappings and perhaps slight shifting ultimately necessary will take a very long time indeed.

When first laying down the plates, tacks are used for temporarily fastening to the block, so that register may be made conveniently, but when everything is satisfactory in this respect small screws shoulα ʋe substituted. Unless the underlay is put under the wood, instead of immediately beneath the electro, the frequent removal of the tack is a matter of necessity. When the underlaying and making register are completed, however, a few tacks or pins may be put in at various points to insure the plate lying perfectly flat to its mount. The heads of the screws should be countersunk, which may be done by means of a small brace and centre-bit, to prevent possibility of blacking. When imposing, due care must be taken with regard to the " pitch," or it may be necessary to cut a piece off the board to allow it to fall in the correct position on the coffin.

Having completed the imposition of the first tint to be worked, lay it on the bed of the machine, measuring from the edge of the ink-table for lay.

The cylinder must be dressed. Blanket, we may say, is rarely if ever employed, the hard-packing system being in every way better adapted to colour printing Of necessity there are many whites in colour formes, and the blanket, as we have had occasion to mention before, is liable to dip, which would cause a hard line at the extremities of the work, besides which the possibilities of dips owing to the detached pieces are largely increased. The cylinder must be prepared in manner described in Chapter XIX. If the forme is fully low, a piece of wrapper may be placed underneath. We are presuming that sufficient time is allowed for underlaying in the proper and best manner, *i.e.,* immediately beneath the plate.

Sometimes—when time is an object, the work of an inferior quality, and the number to be printed very short, necessitating the time employed in making ready to be reduced to a minimum—the sheet is patched in its entirety, and placed under the block. It follows that in this case much thicker paper must be used, otherwise the effect would be practically nil. Under any circumstances this is a slovenly and altogether unsatisfactory way of proceeding, especially if the plates are at all uneven.

Having secured the entire forme in the coffin, roll up in black a sheet on the surface, and pull an impression. It is not necessary to repeat the instructions given in the chapter on making ready. The system of patching the underlay is identical. The greatest care, however, should be taken that the edges are not too sharp, and it will be found advisable to pare them down well in order to subsequently simplify the work on the overlay. Cut the various pieces tolerably close to the work, and lay them on their respective blocks, when they may be fastened with paste in the ordinary way, and the blocks replaced. Strike-on the machine, allowing the cylinder to run so that the underlays may adapt themselves to the plates. A sheet of paper, as usual, must be laid on the surface of the forme.

After pulling another impression, patching the second underlay, and pasting the same in position, the register must be perfected. If we are employed on the second and subsequent colours, the points on the laying-on board must be adjusted to the holes already in the sheet, but in the case of the first colour the lay should be accurately made, and a pair of points or spurs fixed in the gutters

for use in the following forme. The first colour, supposing it to be a tint, will have been made up to the
key, which has already been gauged; and the greatest
care must be exercised in this, as when the initial colour
is printed the succeeding ones will of course have to be
fitted to it. The work of registering is materially simplified when the key is printed, as the slightest deviation is
easily discernible, whereas if several detached tints are
done first, the work is more difficult. In many cases
the artist arranges for the outline to be started first, but
this is obviously impossible in designs full of colour, as
the outline or drawing would be entirely obliterated and
rendered absolutely useless by the opaque colours—
vermilion, blue, or yellow. However, in any case it is
desirable to print a comparatively full block as early
as possible ; but the order in which the colours are
to be printed ꞌwill have been decided originally by the
artist, who will have supplied the blocks to the pressman for proving.

Some printers, when looking through the colour sheets
of a chromo prior to printing, slightly alter the order of
working—that is to say, not strictly following the successive proofs. We think this is a mistake, as immediately
the order is departed from, responsibility is assumed
which does not properly belong to the printer ; in addition
to which there is no combination in the proofs as a
guide. If the original proof is thoroughly satisfactory,
follow it—that is by far the safest plan.

When registering a forme the plates are temporarily
secured by tacks, as we have before explained, in order
that any slight alteration of position may be easily
effected. It is not to be supposed, if ordinary care has
been exercised in the original instance, that any great
amount of shifting will be necessary. The plates may

possibly be put in register by a slight tap in the desired direction, and the removal of one tack and the substitution of another. The difficulty in this matter is considerably increased if the electros are defective. If the plate is a large one, it is frequently found that it is really smaller than the original block, in which case it is obviously impossible to make accurate register, as by tapping in one direction to insure fit, it is thrown out in the other. If the work is very open this difficulty may be overcome by cutting up the plate into pieces and fitting them independently. When the plate is full of work this cannot be done, and another plate must be procured, or the result will of course be faulty.

The electrotyper should, if possible, mould all the colour blocks of a job at the same time, as there is then greater probability of the plates fitting. If, for instance, they are moulded at intervals, the wax of the mould is liable to be of different temperature, and the contraction unequal. If a certain colour is moulded on a cold day, and the succeeding one on a warm, it is very unlikely that they will fit, as in the former case the plates will certainly be slightly smaller, owing to the altered atmospheric conditions. Some electrotypers make a specialty of the production of plates for colour printing, and to these we would recommend the work to be sent.

While on the subject of plates, we would mention the desirability—we may say, the necessity—of having those plates to be worked in vermilion (or in any colour wherein it is used), or blue, to be either brass or steel-faced. With black, electros will last for hundreds of reams, the wear being thoroughly legitimate ; but the mercury of the vermilion and the acid in the blues quickly destroy the surface. In fact, unless the number is extremely short, and reprints are unlikely, the slight extra cost of brass coating as above is a very small matter, compared with the trouble that the plates will inevitably cause if not so treated.

Although there may not be so much work or variety of texture, generally speaking, in the colour blocks, the overlay should be cut with all the care that is, or ought to be, bestowed upon the key or outline. As in the cutting the overlay of an engraving, the press proof should be to hand for reference, and if the colour was originally drawn upon the stone, it is advisable to have the litho impression also, that it may be seen to be exactly what is required.

The chief difficulty with the colour printer, next to securing exact register, is the matching of the tint. We have little to say with reference to overlaying, as we have already dealt with the matter. During the process no doubt many slight difficulties will arise, such as blacking and dipping, especially with formes that are made up with pieces, but experience will soon suggest efficient remedies. Before commencing to patch the first sheet, the formes should be carefully examined to ascertain if every page or piece is, in the first place, perfectly flat to the mount and *well secured.* Injudicious patching, or too much of it, will frequently cause the plate to become springy; this will affect the rolling, and a slur is likely to result, as the plate will have an inclination to move under the impression. An additional screw or pin will sometimes remedy this. After pulling the cylinder and pasting up the overlays, the ink may be put into the ductor and the colour run-up.

Rollers for coloured ink should be slightly harder than those used for black work. Coloured inks, especially body colours, are stiffer, and consequently possess more "lug," than black ink, and therefore the composition must be slightly firmer, or it is liable to become destroyed after a short time. The patent composition is peculiarly suited to coloured inks, as turps, which must be used to effectively remove the colour, is well adapted for cleansing purposes; whereas the ordinary composition is speedily affected by its use, and there is some difficulty in effectually and quickly washing a colour roller by means of ley. We have known rollers of patent composition to work upwards of 1,000 reams of colour, and still be in a fit and proper condition. Every machine-minder will admit, however, that the life of a roller, be it made of ordinary glue-and-treacle or the patent composition, is very much dependent upon its being properly treated. Success is largely owing to the proper condition of rollers, and they should experience at the minder's hands, or at the hands of the boys under his immediate supervision, careful treatment. Rollers have much to answer for, we are aware, but very often they are rather sinned against than sinning, as every colour printer of experience will testify.

With reference to the ink, the name of the special colour or colours employed should always be stated on the pressman's proof sheet as a guide to the machine-minder. It is seldom that the ink is made up entirely of one colour—usually by a combination of two or more. This is determined in the first instance by the artist when proving, and the pressman should carefully note the various proportions, otherwise some difficulty may be experienced in obtaining the exact shade. With ordinary showcards or broadsides the stock primary colours are invariably used, and as printers as a rule keep in stock cans of ink—say vermilion, yellow, blue—supplied by the maker in the same shade time after time, no difficulty is experienced with this class of work. But with chromo plates there is perhaps hardly one tint which is produced from a single colour, and inasmuch as a particular shade may frequently be approached by entirely different combinations, it is necessary that a record be kept, when proving, of the various pigments used, together with the amount of thin, middle, or strong varnish employed.

The colour printer often mixes his own colour, and therefore it is highly desirable that he be thoroughly cognisant of the various peculiarities of the pigments or inks. He knows, or soon learns from experience, the. proper materials to employ to obtain a given shade, even without the friendly notes on the press proof ; and also the condition that a roller should be in to most successfully distribute and evenly deposit the colour mixed. The ordinary machine-minder who is at work constantly on book-work simply deals with the ink as supplied by the maker, and therefore his work is to a great extent simpler than that of one who has not only to be constantly on the alert to see that the register is perfect, but also that the work is good and the colour even and correct.

We may here refer to the practice (growing of late) of chromo printers making their own inks. Where the establishment is small, it is perhaps questionable economy. To be thoroughly successful, the service of an experienced man is necessary, and when we consider that the wages amount to, say, 35s. to 40s. per week, added to the cost of the plant that is necessary to carry out the matter satisfactorily, it is obvious that a large amount of ink must be ground to make it pay. Not that it is at all necessary that all the ink required should be sent in from the makers exactly in a fit state for putting into the ductor. This would of course be almost impossible, especially in the case of tints, which the minder generally prefers to make or mix himself, as usually varnish constitutes the largest portion. The ink of ordinary consistency is sent in, and reduced by thin or "mid" varnish to the necessary shade.

In lieu of mills, stone mullers are sometimes employed wherewith to grind up the pigment, but this is only advisable when the quantity of ink is small and the colour soft. With umbers and such gritty and stubborn colours this method is simply impossible. If the quantity of ink habitually required is such as to warrant a special department, a light and airy room with a stone floor should be selected, and, if

Fig. 109.—Muller and Stone.

possible, shut off by iron doors from the main building, as the smells arising from the mills, especially in the case of blues, are far from pleasant or healthy. Two stone rolls should be provided—one for the dark colours and the other for yellows and reds. Granite is to be preferred, as steel is apt to destroy the delicate colour, or that "bloom" which is or should be characteristic of good ink.

One set of rolls should be driven at a comparatively low rate of speed, otherwise the heat generated from the friction will destroy the lustre, especially in the case of blues.

Vermilion is the simplest of all to grind; in fact, it hardly requires grinding, only thoroughly mixing, as it is supplied in fine powder. Ultramarine, again, is another pigment which is comparatively easy to deal with, and chromes, although supplied in lumps, are extremely soft, and consequently cause little trouble in the mixing. With burnt sienna, umbers, and such hard pigments, how-

ever, it is altogether a different matter, and the greatest care and experience are necessary on the part of the maker to render them fit for the machine. It is a matter of time and patience to produce ink that will work freely. Often, notwithstanding it has been put through the rolls

Fig. 110.—Ink Mill.

five or six times, it will be found to be gritty on the ink-table, and to fill up all the fine work on the forme.

Bronze-blues, again, require the utmost care in the making, otherwise the essential charm of the colour, the bloom, is entirely lost, and the result is a dirty black. When, therefore, we consider the difficulties attendant upon the production of many inks, it is questionable, as we have before said, whether ink-making is profitable, always excepting that a continuous large consumption justifies the employment of an experienced man, and the laying-out of a considerable amount of capital for the necessary appliances.

In calculating the cost of ink-making, the necessary

extra driving power must be taken into consideration. An ink mill consists of three rolls, each travelling at a different speed. It will therefore be understood that a large amount of friction is existent, which absorbs power. If the pigment, such as vermilion, merely requires *mixing*, it is only necessary to turn the rolls ; but when absolute grinding or pulverising has to be done, the power required is proportionately greater.

With reference to the purchasing of dry colours, it is extremely inadvisable to adopt the cheapest. Both chromes and blues may be obtained at 8d. or 9d. per lb. ; but, as a rule, they contain an excess of acid, which speedily destroys the plates. All colours should be well washed, and this process costs almost as much as the pigment. Pure chromes and superior blues should always be selected, as they will be found the most economical.

With vermilion it may be said, as a rule, that the quality itself varies little. There are different shades, the extra-pale being generally preferred. Sometimes, however, the cheaper qualities are adulterated with a material which adds to the weight.

Having dealt in a brief manner with the ink, we have again to refer to the starting of our forme. As with ordinary black work, a second overlay will be required to insure the quality of the work being equal to the guide proof. After the impression and register are satisfactory, the setting of the colour may cause some difficulty, and the capability of the machine-minder to speedily and accurately fix the exact shade is a distinct and almost unvarying test of thorough efficiency.

The general directions with reference to the preparation of one forme of course hold good for the following ones. But difficulties arise frequently when the final colours are being printed, owing, sometimes, to the body of colour already on the sheet failing to pick up. This is often consequent on the surface being either too hard or too soft. If the initial lines drag up with a glossy surface, there is a difficulty in making the ink pick with the desired uniformity. This defect is often attributable to the hardness of the paper, which will not allow the varnish to sink. Sometimes this may be obviated by softening the ink, but even if this be done, there is the danger of an ugly set-off. Talc may be brushed over the sheet, but this is obviously impossible with long numbers, and its effect is always detrimental, as it destroys all the brilliancy of the work.

The previous tint should be fairly dry before the next is put on, but at the same time, if it becomes thoroughly hard, the difficulty of picking up will again arise. When heavy body-colours are printed, set-off paper should be used. This is especially necessary if a back has to be printed ; but, under any circumstances, it is desirable to interleave after the third or fourth working, for besides the certainty of the back of each sheet becoming dirty, there is a probability of the heaviest parts adhering to the next sheet, and if this occurs an immense amount of trouble is in prospect. Whether the work be interleaved or not, it should never be piled in too large stacks, but laid out in small lots, that the pressure on the newly printed work may be reduced to a minimum. A good plan is to have special boards made, on the ends of which at right angles are fixed solid pieces of wood, about six inches deep. About half a ream of paper may be put between the upright pieces, and then another board placed on top, so that a number of lots may be piled on each other, without any of the work being subjected to undue pressure.

The dirty backs of presentation plates are often

remarked upon. In many cases the number printed is so enormous—taking the *Graphic* and *Illustrated London News* for instance—that systematic interleaving would be almost impossible. But, at the same time, although simply a soiled back does not materially depreciate the value of a chromo plate, the fact remains—that the colour is deposited on the back at the expense of the pictures themselves—a fact which is at times painfully apparent.

The printing of showcards or heavy broadsides, whereon a heavy body of ink is used, is necessarily a slower process than that of producing the ordinary chromo plate, as it is impossible to deposit on the plates, or lift a mass of colour, very rapidly. The speed must be regulated by the nature of the forme and the stiffness of the ink used. If a large body has to be covered, the machine should not be allowed to run at the rate of more than, say, 250 copies per hour. This may appear to be an exceedingly slow process, but it is really the only one by which the heaviest class of work may be satisfactorily done. If it is imperative that a quicker speed be employed, the body of ink must of necessity be reduced, otherwise the paper would be torn from the grippers. The secret of successful broadside work is good paper and stiff ink. Thin or poor paper will not hold the necessary body of colour.

On the Bremner and also on the *Graphic* machines is fitted a supplementary adjustable cam, which may be made to clamp the fly-wheel when the impression is being taken—easing the lifting. This is a great convenience to the colour printer.

Broadside printing is generally looked upon as being, if not the commonest class of colour printing, certainly by no means entitled to much consideration. This is a mistake. We are free to admit that the majority of the bills which adorn London hoardings are far from being excellent examples. But in many cases the care necessary, and the excellence of the material used to produce a striking bill, would surprise many printers. When it is stated that one firm, who make high-class bills a specialty, dry each sheet singly between the workings, employ only the best pigments obtainable, and print on plate paper, equivalent to say about 80 lb. double demy, it will be admitted that broadside printing sometimes receives more care and attention than is generally supposed. Of course this kind of material is costly, but when it is considered that the bills thus produced last for a very long period, and retain their brilliancy under prolonged exposure, they are really economical.

Both pine and sycamore wood are used for cutting broadside work. The former is cheaper than the latter, and is much easier to cut, but the latter is to be preferred, as it is much harder, and there is an absence of grain, which sometimes becomes prominent in impressions from pine blocks. Sycamore also really takes less colour, as it is not so absorbent.

Machines constructed on the Wharfedale principle are to be preferred for colour printing. The Bremner, *Graphic*, and Dawson being especially suitable—the rolling power in each case being ample. Those provided with double inking gear—*i.e.*, ductors, and the duplicate set of rollers at the laying-on end—are necessarily the best, as both the distribution and deposition of the ink are superior. It is advisable that an extra girder under the cylinder, to insure absolute rigidity, be fitted, as when full-sized tint blocks are printed an immense amount of power is required in order to secure a faultless impression.

Platen machines were at one time very largely used for colours, but only in very exceptional cases are they now adopted—in fact, only in houses where there is an indisposition to abandon old plant. They take only a comparatively small sheet, the inking is not so good as in the Wharfedale, the speed is low, while no advantages can be claimed in facilities for superior register.

In laying-on very large and heavy sheets, it is necessary to employ two boys to point—one on either side of the board. With smaller sheets, however, the point holes may be placed within easy reaching position of the boy—one on the near-side and the other at the centre of the top edge of the sheet.

Bronze Printing may be characterised as tedious, expensive, and unhealthy work. When possible, this is always done first, as, should the powder be dusted upon a sheet upon which more than its own colour is printed, there is the liability to "tint."

Pale chrome is used for the groundwork, and the bronze deposited by pads made of cotton-wool, immediately the sheet is printed. A large board, covered with glaze-boards, and having a flange round the edges, say two inches deep, is the best for the purpose. The number printed by the machine is necessarily limited when the dusting is done by hand, as, unless a number of boys are employed, it is impossible to bronze as quickly as the machine will print. The sheet is lifted immediately from the cylinder, and taken to the board before mentioned, when the bronze is laid on and rubbed lightly over the entire surface. The finishing process of dusting-off can be done at leisure.

The bronze should be placed in small receptacles, and the pad lightly dipped on to the surface, allowing a sufficient quantity to adhere to cover the sheet. Although the work must be ultimately brushed or wiped again, the lad should be careful not to allow superfluous powder to lie on the surface, but should clean the sheet as far as possible. It must not be rubbed too hard, or the ink will be spread, and the powder will attach itself to the excrescence.

When possible, the bronzing should be done in a separate room. We know this is often inconvenient; but the powder, being very fine, flies off and covers everything around. Added to this, it is inhaled by the lungs of the workmen, and causes an irritation which may result in a serious illness. It is the custom in the majority of houses to allow the men and boys a quart of milk per day, which to a great extent nullifies the pernicious effects of the powder.

When the work is dry after bronzing, the sheets are rubbed singly with a clean rag, which removes the superfluous powder, and tends to impart a brilliancy. Greater force may now be used, as the ink is hard, and there is no danger of its spreading.

When there is a quantity of bronzing work, a machine should be employed. With half the number of hands at least double the quantity of work may be done, added to which the bronze is economised, being confined within the machine. About 250 to 500 copies per hour may be bronzed, depending upon the size of the sheet. If the printing is performed at this speed, the taker-off has time to carry the newly printed sheet to the bronzing machine, should it be in proximity, as is desirable.

Bronzing machines of recent manufacture are both compact in appearance, occupying little space, and effective in the working. There are several kinds before the trade, but all may be said to be constructed on the same principle, although varying considerably in detail.

The bronze is placed in a narrow trough extending across the machine, and is sifted on to the surface of the sheet as the latter passes round the cylinder. Revolving brushes remove the superfluous bronze, and the sheet is delivered upon a slanting taking-off board, no boy being necessary. As before mentioned, the whole of the machine is well covered in, so that no powder can escape. The waste material which is finally brushed off the sheet is accumulated in a special receptacle at the bottom of the machine.

Perhaps the best known bronzing machines are those manufactured by Leeming, Ray, and Co., and W. B. Silverlock.

After the work is dry, it may again be passed through the bronzing machine, previously having removed the powder. This second process will finish the sheet.

Sometimes it is necessary to print the bronze after other colours have been worked. This is the case when tint grounds are employed, as it would, in many instances, be impossible to register the subsequent general colours to fine lines or ornaments in bronze. There should be little difficulty in this if due precaution has been taken that the inks of the previous colours are *perfectly dry*. If not, the powder will cling to the soft parts. It is a good plan to roll the sheet before bronzing, as this tends to smooth the ink already on, and take out any rough places that would hold the powder. Coarse bronze, or brocade, is the best to use under these circumstances, as it is not so liable to " tint."

With reference to the purchasing of bronze powders, it may be stated that the chief difference between the cheap and the moderately expensive material is the comparative fineness, and the special care taken in the

A A

finishing process—the extracting of grease, &c. They are all manufactured of the same class of metal. Most of the powders come from Germany. They may be bought for 2s. 3d. per lb., but although we do not advocate quite such a cheap bronze, it is really unnecessary to pay the excessive charges which were common some years ago. Bronze may be purchased at 32s. per lb., but it is a mistake to suppose it is as much better than the cheaper qualities as the difference in price would suggest.

There can be little doubt that bronzes will change colour when exposed for a time. The fact of their doing so is no actual proof of the inferiority of the material employed. Atmospheric conditions enter largely into this matter, as in a manufacturing town a showcard in gold will very soon be turned positively black, owing to the various gases, &c., given off from the factory chimneys.

The same remarks apply to vermilion. This pigment, when chemically pure, is practically permanent under normal conditions. But the sulphur from the smoke will, in time, attach itself to the surfaces, and form a positive black. This, however, is not the fault of the vermilion. If the black surface be carefully scraped, the colour underneath will be as bright as ever.

Coloured enamel papers sometimes cause the bronze to become discoloured even when covered up. This may be attributed to the action of the pigment used for colouring the enamel.

CHAPTER XXV.

SMALL JOBBING MACHINES.

The Cropper — Universal — Bremner — Mitre — Caxton — Liberty — Gordon — Model — Empress — Perfection — Golding Jobber — Godfrey's Gripper—Greenwood and Batley's Rotary.

FOR small commercial work the treadle jobbing machine is generally adopted, and with careful boys this may be considered one of the most profitable investments in the printing office. Small Wharfedales are called jobbing machines, but we are at present referring to that class originally known by the common name of the Cropper.

The invention of an American named Gordon, they first became generally known in this country in 1862, when they were exhibited, and obtained medals, at the London Exhibition of that year. Although there are at least twenty firms employed at the present time in their manufacture, they are all built on the same general principles. The first of English make, named the Minerva—better known as the Cropper, after the Nottingham firm of engineers—met with immediate success, and a very large number were sold, many of which are still doing good work in almost every city throughout the country. Very soon, however, competitors entered the field, and almost every year, with various claimed improvements, others were offered to the trade. The principle upon which they were originally constructed, however, seems to have been so sound, that after the lapse of twenty-five years, very little alteration is apparent in the general appearance of the machine. The inking

A A 2

apparatus has certainly undergone, in some cases, a modification, and the platens have been strengthened, but practically these improvements are all that can be claimed by
the recent makers.

Fig. 111.—The Minerva, or Cropper.

The whole of the working parts are contained between two slight frames, the inking apparatus being at
the top.   The forme is fitted perpendicularly into a bed
in front of the layer-on, the special chase being bevelled
at the side, and secured at the top by a small movable
bracket.   In front of the bed is the platen, lying at an
angle of about 45° when stationary.   The sheet is laid
to pins on the platen, which is pulled forward towards

the forme by two strong steel arms, one on either side ; returning to its original position, when it is stationary sufficiently long to admit of the boy taking off the printed sheet with his left hand and substituting a clean one with the right.

Upon the extremity of the driving shaft, outside the frame, is a small spur wheel, which works a large cog. To this latter wheel is attached a strong steel rod, fixed at the other end to a frame, working loosely upon fixed bearings. The rollers, generally three in number, are fitted into their forks on a level with the ink disc or table. By this method the rocking frame moves backwards and forwards, as the large wheel, holding the steel arm, revolves. It will be noticed, however, that the rollers, after leaving the ink table, must necessarily continue their travel almost at right angles, to ink the forme, which is in the perpendicular bed. This is managed by strong springs attached to the forks, which pull the inkers flush with the surface of the forme, allowing them to run again on to the ink table on their return. The rollers are fitted with flat wheels, and run on guides up and down on either side of the bed, so as to prevent the composition from being cut or damaged by undue pressure, and at the same time insuring steady motion and support.

The ductor is placed at the extremity of the travel of the last inker, which touches the steel roller on every return, taking the regulated quantity of ink. The ink table varies in make. The original Cropper is provided with a disc, which moves slightly round between every impression, thus facilitating distribution. This was afterwards supplemented by two discs, one working flush in the other, but moving in an opposite direction. Some machines have square tables with revolving disc in centre, while in others the disc is discarded altogether, and a solid table substituted.

The forme and platen are brought together by the strong steel rods on either side, one extremity being attached to a disc slightly below the bed of platen, and the other to the frame at back of type-bed. The motion is communicated by a cog, working in gear one side with the driving shaft ; on one extremity is fixed a large fly-wheel, which assists the impression.

It may be remarked that nearly the whole of these machines offered to the trade are excellent. The original ones were so good, that it is obvious that if any of the recent make were in any way defective, they would be immediately condemned. All are fitted with striking-off gear, thus enabling the boy to save a badly laid sheet.

They may be safely run at 1,000 copies per hour, but this speed is somewhat difficult to maintain by treadle. It is advisable to have them driven by a belt, when better work and greater quantity may be relied upon. When a boy has to work the treadle, lay on, and take off, it can hardly be expected that the work will be so well or so rapidly performed, as his attention and energies are divided, and variations will necessarily occur in the speed. In the case of solid formes the work is very heavy. All machines are supplied with riggers, and the application of steam-power is a trifling expense compared with the advantages.

Making ready on these machines takes but a short time. The sheets are secured on the platen by a slight movable frame. After the forme is placed in position, it should be underlaid in the ordinary way of plates. Prior to patching for the overlay, pull an impression on the sheet on the platen, patch, and place in position.

The formes, being small, are soon prepared, and an ordinary commercial job should not take more than half an hour in the starting. It is advisable to use the patent composition rollers; they are less liable to become injured by the somewhat rapid travelling.

By judicious adjustment of the impression, light rule-work or heavy cut-work may be equally well done. With careful laying-on, excellent register work can also be turned out.

In the foregoing we have referred to the jobbing platens as a whole. It would be invidious on our part to select any one as being better than the others. The question of price must necessarily enter largely into the matter; and if strict economy is to be studied in the original outlay, it is, of course, unreasonable to expect such strong and well-fitted machines as when a much higher figure is paid. Not that it should be supposed that the low-priced machines are inferior. They are certainly lighter in construction, but, as a rule, are so well balanced that they invariably give satisfaction. But extra labour in the cutting of the cogs, and the use of best steel, must necessarily add to the cost and value of a machine. The following are a few of the jobbing platens now used, and although, as we have before pointed out, they are made on the same lines, each possesses some distinctive feature, which, in the opinion of the respective engineers, of course renders it equal or superior to others.

The Universal is a machine suitable to heavy work, the inking arrangements being particularly good. The forme is placed in a stationary bed. The platen, which is of very solid description, is provided with six regulating screws underneath. The inking apparatus consists of a series of revolving iron cylinders at the back, which are supplied with ink from a small ductor at the base. During the time the platen is stationary, when the sheet is laid on, the inkers are travelling round, contributing to perfect distribution. The frame working the inkers

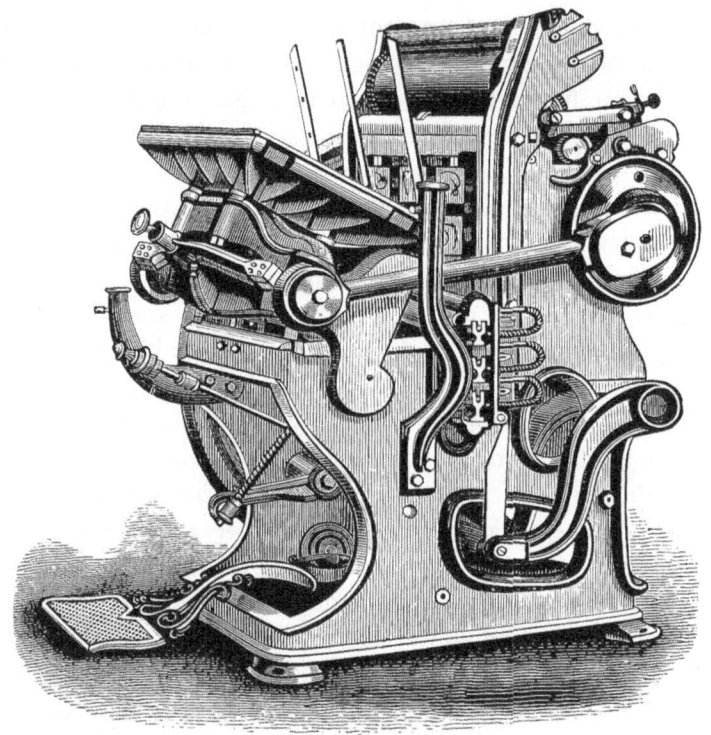

Fig. 112.—The Universal.

is worked by a cam inside the large geared wheel, the shaft of which is attached to steel bars, which brings the platen to the bed.

With a careful layer-on perfect register may be obtained, and we have seen work turned out in three and four printings which would have done credit to a litho machine. Heavy wood blocks, with sound making ready, may be printed in first-class style, both the distributing of the ink and the impression being excellent.

Fig. 113.—The Mitre.

The above remarks apply more particularly to those machines of later construction.

Harrild's Fine Art Bremner Platen is singular, perhaps, inasmuch as larger space is devoted to the inking. A series of wavers, fixed at angles in a travelling rame, are provided. These are independent of the inkers, and, like the distributing rollers of the ordinary printing machine, do not leave the surface of the ink table. The machine may be stopped or started immediately by the action of the foot, the operator's hands being thus at liberty. As is characteristic of all Messrs. Harrild's work, the fitting is in every respect excellent, and numerous small details are added, which greatly facilitate both the management and running of the machine.

The Mitre.—A very substantially constructed machine, made by Messrs. Dawson, of Otley. The inking arrangements are somewhat similar to the Universal. Although very solid, the platen is very carefully and evenly balanced, and the power required for driving is comparatively small. To avoid slurring, the platen is guided strictly in a parallel position to the face of the forme. The heaviest cut and embossing work may be safely entrusted to this machine.

The Caxton.—Made by Messrs. Furnival, of Reddish. This machine is excellently constructed and well balanced, rendering it very easy to drive. The duct roller, which takes the ink from the ductor, passes entirely over the disc between each impression, thus aiding the distribution. The rollers have a long travel, the last roller passing entirely over the extreme end of the forme. The chase is self-fitting, and requires no fastening when forced into position. The "dwell" on the impression is also prolonged.

Among others of this class we must not omit to mention

The Liberty, an excellent machine, and among the first offered to the trade. The inking arrangements vary from other treadles, inasmuch as the disc or inking slab is placed at the back of the machines. It is light in build, and may be safely run at a high rate of speed.

Powell's Improved Gordon.—Singular, perhaps, in

possessing an extra-large inking surface. Three new motions have also been added to the later machines—stop-platen, roller-lift, and double inking.

The Model, originally known as a small press, was

Fig. 114.—The Liberty.

rather constructed for amateurs than for the trade. The makers ultimately enlarged and strengthened the machine, and they now receive a fair measure of patronage, certainly justified by the excellent manner in which their machines are built.

The frame is in one piece, the bed being stationary. The platen is forced to the forme by knuckle-joints attached to a short arm between two cog-wheels driven

direct from the main shaft.   Impression screws (five in
number) are fitted under the platen, by which means
the pressure may be regulated to a nicety.   A patent
apparatus is placed at the side, to throw off the

Fig. 115.—The Model.

impression at any moment when desired.   The inking
arrangements are very similar to those of other platen
machines.   The inkers, however, are provided with a
small brass rider.

As will be noticed on reference to Fig. 115, this
machine is constructed without side-arms.   The makers

claim this as a distinct advantage, as a larger-sized sheet than the machine can print may be laid on without the possibility of the edges that overlap on either side becoming damaged.

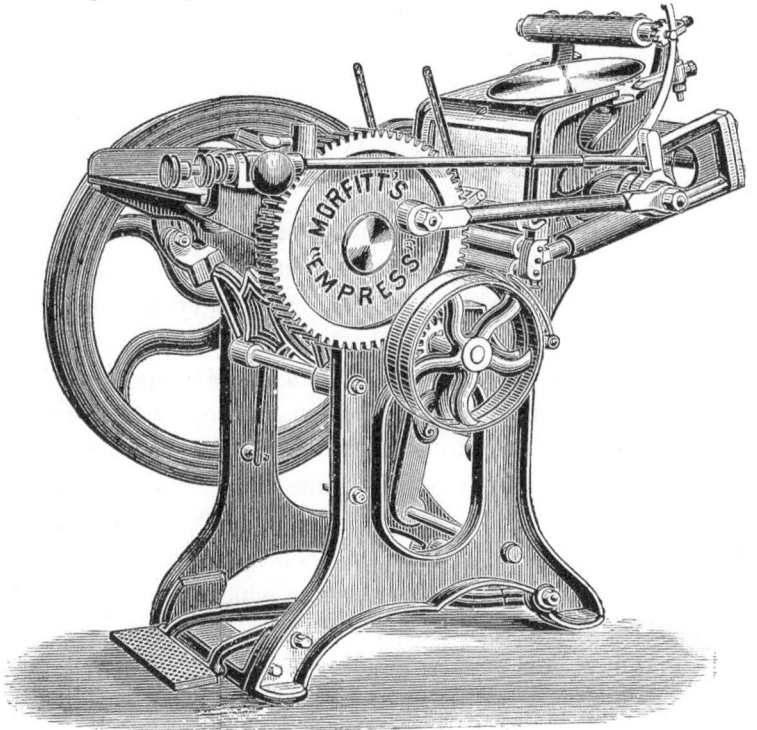

Fig. 116.—The Empress.

All the parts are interchangeable, and thus a break may be speedily repaired. All the castings are kept in stock, and may be substituted by the printer himself, obviating the necessity of employing a mechanic.

These machines are sold by the Model Press Company, and are both economical and reliable.

The Empress, very similar to the Cropper, is of light construction, and well adapted for commercial work.

The bed moves slightly forward to meet the platen.
Like the Cropper, the Empress is to be recommended
when it is desired to drive by treadle.

The *Golding Jobber*, a recently introduced American
platen, is slightly different in its construction from those
before mentioned. The gearing and working parts are
placed below the bed and platen. Like the Mitre, the
frame and bed are in one casting. The manufacturers
claim that inasmuch as "all the movements are positive,
without cams, eccentrics, or slides," there is no possibility
of back-lash or slur. The impression is obtained by a
compound movement, by which means the power required
to drive is very small. The inking arrangements are both
novel and effective.

Godfrey's Patent Gripper Platen Machine possesses an
ingenious combination of the platen and rotary motions.

The forme is placed on the coffin of the revolving
cylinder. The platen occupies the same position as on
the ordinary jobbing machines, but the lay is materially
simplified by an automatic arrangement—the paper being
placed on a stationary board, from which it is taken and
adjusted to its exact position.

The layer-on has really only one duty to attend to,
the taking-off being accomplished by the machine itself.
The boy has therefore both hands at liberty. There is,
moreover, no danger of the hands becoming injured, as the
sheet is laid on a board, as mentioned above, and not on
a movable frame. The inking arrangements are excellent,
and the impression is absolutely dead.

The number that may be printed is really only
limited by the speed at which the layer-on can feed-in
the sheets. This machine is essentially suitable to printers
who have a constant supply of small jobbing work of
large numbers. It is manufactured by Messrs. Furnival.

FIG. 117.—GODFREY'S PATENT GRIPPER PLATEN MACHINE.

Messrs. Greenwood and Batley, of Leeds, have adapted the rotary motion to the jobbing machine. The reel of paper is slung into position on the top by a small crane.

Fig. 118.—Greenwood and Batley's Rotary Jobbing Machine.

This machine possesses adjustable knives, that the work may be delivered cut up to the desired size. For long numbers of ordinary commercial work it is eminently adapted.

CHAPTER XXVI.

THE WAREHOUSE.

Warehouse Machinery and Appliances—Drying and Pressing—
Hydraulic Presses and Pumps—Boomer and Boschert Press—
Cutting Machines—Gill's Hot-rolling Machine—Perforating and
Eyeletting Machines, &c.

THE warehouse proper is, generally speaking, the department into which the newly printed work is carried,
dried, pressed, and packed for delivery to the customer
or binder, as the case may be. The constitution of this
department of the printing office is necessarily dependent
upon the special class of work done in the establishment.
In houses where first-class bookwork is printed, proper
arrangements for drying and pressing are, of course, necessary. But where commercial work, and work of a
miscellaneous description, are turned out, various minor
machines, for perforating, eyeletting, scoring, numbering,
ruling, card-cutting, &c., are also required.

Immediately a ream of work or a small job is
completed, it should be taken to the counter to be
checked, that the exact number of sheets may be ascertained prior to the job being lifted from off the machine.
A full ream of paper delivered from the wetting-room
should consist of 516 sheets, including top and bottom.
Practically, therefore, there are 514 good sheets. Allowing six for spoilage, only 508 remain, and this quantity
should be sent to the counter ; otherwise, especially with
small numbers, the total quantity, when pressed, rolled,
&c., will possibly be short, as a few sheets may be spoiled

B B

in the warehousing, and the binder also requires a few over in case of accidents.

In large establishments the counting-room is separate from the main warehouse, although practically under the control of the same foreman.   It should be in close proximity to the drying-room, into which the work is carried after being counted.   The drying process varies in system with almost every house.   The best and most inexpensive plan, perhaps, is to employ a large, lofty room for the purpose, having brick walls and an iron door. Within a short space of the roof should be fixed iron rods, supporting a quantity of racks—formed by two parallel bars hanging down, having stout wooden laths extending from side to side at intervals.   The tops of the iron supporting rods should be hooked, that they may run loosely, and be shifted by means of the peel to any desired position.   Steam pipes connected with the exhaust may be arranged round the room, and by this means a temperature of, say, 100° may be obtained. The sheets should be hung in small sections across the bars.   If too many are placed together they hang closely, and the hot air is unable to pass freely between them, and as a consequence the drying process is uneven.

All gas-lights should be well protected, as the paper is of course very inflammable under these conditions.

After hanging for twelve hours, the work may be in a fit condition to press.   This will, however, depend upon the amount of ink used, and also upon the condition of the paper.   It is never advisable to allow the temperature to exceed 110°, as excessive heat is apt to turn the ink brown.

Superfine cut-work should be dried by cold air.   This is a slow process, and demands considerable space, but by this method the ink will retain all its original brilliancy, which frequently is destroyed when heat is employed.

The work, when perfectly dry, is fit for pressing. Before laying between the glazeboards, its fitness should be ascertained by laying a piece of white paper upon the most solid part, and slightly rubbing it with the thumbnail. If the paper is soiled, the ink will set-off on to the glazeboard under pressure, and is therefore not in condition to be put in the hydraulic. Hang it up again, or the glazeboards will have to be well cleaned after every pressful, and the quality of the work will suffer.

The HYDRAULIC PRESS consists of a heavy head-piece supported at each corner by iron pillars. At the base is a solid plate or bed, to which is attached a ram, working tightly in the collar of the cylinder The water is forced into an opening at the base of the cylinder by means of a solid plunger, through specially constructed pipes. When the water is injected and the plunger returns, previous to forcing in a further quantity, a self-acting valve closes in the water-pipe, and prevents the weight of the ram from returning the water, the valve again opening when outward pressure is applied. When the bed of the press is raised sufficiently high, another valve is screwed down to prevent any possibility of the water escaping. If it is desired to release the work, the valves are opened, and the bed-plate and ram by their own weight force the water back into the reservoir.

A safety-valve is fixed into the feed pipe, the levers attached thereto being weighted. The weights may be placed in any position along the levers. If the intended pressure is exceeded, the water escapes through this valve, and prevents any damage.

B B 2

As an enormous pressure is exerted, it is necessary that the joint or collar hrough which the ram works be absolutely watertight. This is made of a disc of leather, having a hole sufficiently large to admit the ram. The

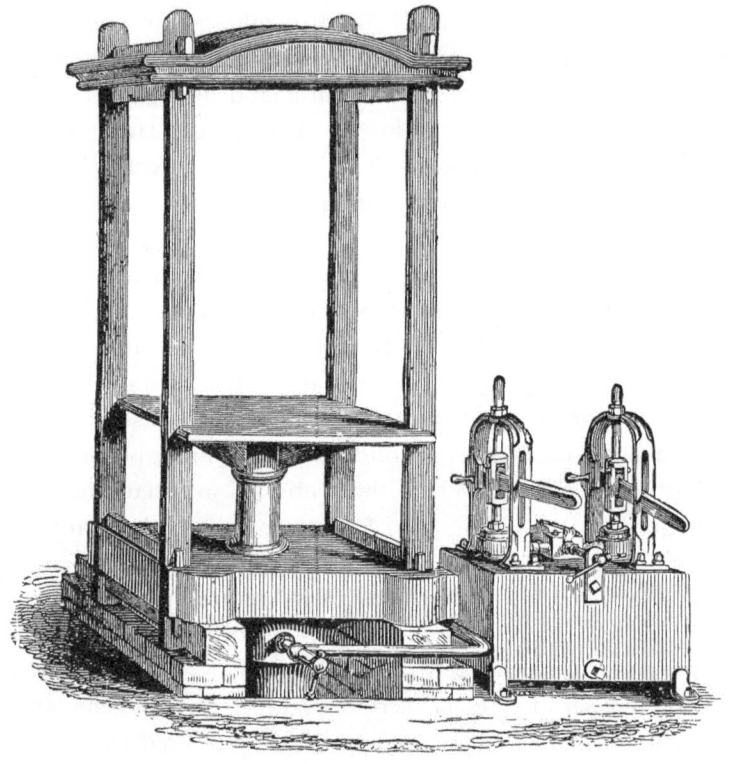

Fig. 119.—Hydraulic Press.

leather is doubled over in a semicircular manner, the concave part downwards. When the water is forced in, the pressure is exerted on the inside of the leather, forcing one side to the surface of the cylinder and the other to the ram equally, and this forms a perfectly watertight joint.

These leathers require to be renewed at times, for

if they become in any way perished, or defective, the water will escape at the collar of the cylinder, and pressure will gradually lessen.

The distance between the bed-plate or table and the head-piece is about five feet, thus allowing a considerable quantity of work to be placed between. The pumps are sometimes worked by hand—very long handles being used to insure greater leverage to give the final nip. When pressing is done to any extent, it is advisable to employ steam. The simplest and best method is to employ a counter-shaft, fixed immediately above the pumps, with loose handles working in eccentrics. The levers, having an up-and-down motion, are secured to the pump-handles by the insertion of strong bolts in corresponding holes. When this plan is adopted, it is necessary to have gauges fixed by the side of each press connected with the pipes, that the exact pressure may be accurately ascertained, otherwise the regulated pressure may be exceeded and some part fractured—most probably the head of the press, which will be split in two. When pumping by hand, and without a gauge, the amount of pressure can be very nearly guessed by the degree of resistance. But hydraulic gauges are not expensive, and by their use pressure may be ascertained to a nicety. It is desirable at all times to have the exact pressure indicated. The necessity for this is apparent when it is known that the work will give in a short time, under great pressure, after pumping up, and it is usual to again put on the pumps, to reach again the regulated squeeze.

When there are a number of presses constantly in work, it will be found that the pump worked from the shafting is much quicker and more economical, as pipes may extend from the same reservoir round the entire series, supplying all. The water may be allowed to run into any press by simply opening its screw valve, and when the gauge denotes the maximum pressure, by screwing down, and shutting off from the main pipe. In the fixing of these, the service of an experienced hydraulic engineer should always be employed, as it is necessary to use pipes specially made for the purpose, and tested to resist a given amount of pressure.

Hydraulic presses should be in every case provided with a substantial and unyielding foundation. Sometimes brick tiers flanged with stone are erected. If this is inconvenient owing to the distance of the warehouse from the ground floor, substantial iron girders may be fixed into the wall below the floor, sufficiently long to allow the base of the press to rest upon them. Although these reservoirs are of great thickness, excessive pressure will sometimes split them.

It is always desirable that a given set of glazeboards should be used for a corresponding size of paper. If a sheet of smaller dimensions is placed in them, the board is apt to become marked at the extremity of the paper, and may, in turn, affect the next batch of full-sized sheets put between. The placing of small work in large presses is dangerous, as, unless the heap be exactly in the centre, an undue strain is liable to be exerted on some part. It is advisable to have the press, say, two-thirds full between the perpendiculars when pumped up, so that the ram may be well supported. Heavy blocks of wood are frequently used to fill up a space between the top board and the head, when only a small pile of sheets is required to be pressed. When this expedient is adopted, every precaution must be taken that the pieces are of exactly the same thickness, or uneven pressure will result. A thick wetting board, slightly larger than the sheets, should be placed immediately under the blocks of wood.

With reference to the amount of pressure, it may be stated that from 180 lbs. to 240 lbs. per square inch may be safely allowed. It should be understood, however, that it is absolutely necessary that the work be thoroughly dry, or the ink will adhere to the boards under this pressure.

Standing or screw presses are sometimes used in lieu of hydraulic, but afford a very poor substitute, as it is almost impossible to take out the marks of impression in well-dried work, especially if the paper is hard and tinny. To overcome this difficulty, it is frequently the practice to put the work in in a semi-dry state, when it is of course spongy and soft ; but although the sheets may appear perfectly smooth when run out after the nip, the impression marks will reappear as the work becomes dry, and present the appearance of never having been pressed at all.

Screw presses are made of various patterns, those of earlier construction having the platen attached to a thick vertical screw, piercing the head of the press. This is brought down at first by turning a small handle at the base of the screw, the final nip being administered by means of a long iron lever or rod, being placed through a hole in the shaft immediately above the platen, and forced round until it is impossible to screw tighter. A more powerful press is that made with a substantial bevel-wheel at the base of the upright screw, working at right angles in a small cog fixed underneath. The shaft upon which the latter works is carried out between the perpendiculars, and the screwing-up motion is perpendicular, instead of horizontal, and greater manual force can be exerted.

For small jobbing work, the ordinary screw presses

are useful, but for large book-work they cannot be commended.

By far the most successful and powerful of screw

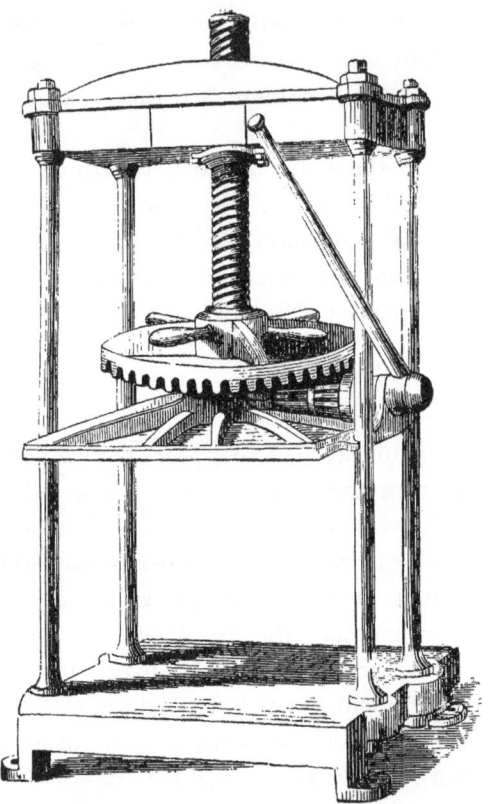

Fig. 120.—Screw or Standing Press.

presses is the BOOMER AND BOSCHERT, introduced from America some years ago. By an ingenious application of the knuckle-joint and screw, a pressure only second to the hydraulic may be obtained. The platen is attached to the head by a pair of knuckle-joints, working on a large horizontal screw. By rotating the screw at

one side, the levers are slowly forced into a perpendicular position, and by the combination of these two
powerful mechanical motions, a maximum of pressure is
obtained with the expenditure of a comparatively small
amount of power. As with the ordinary screw press, a
long iron rod is used to give the final "nip." The
Boomer press is also made for steam power. In the extended screw at the side is fixed a pair of riggers, and
by this means pressure can be exerted fully to the
extent of the hydraulic. This press has not been
adopted so largely by printers in this country as its
merits warrant. Probably the original cost has something to do with this, but there can be little doubt that
it is eminently adapted to the pressing of paper, and
where it is inconvenient to erect hydraulics it will be
found a most effective substitute.

Sometimes two or more sheets are placed between
the glazeboards. The advisability of putting in more
than one must be a matter to be determined by the
warehouseman. Work that has been pressed singly will
necessarily look better than when several sheets have
been put between a pair of boards. For ordinary work,
however, if the ink is thoroughly dry, two to four sheets
may safely be placed together. But if any degree of
finish is desired, a larger number should not be put in.
Quire-pressing is a slovenly expedient, and flattens the
work rather than presses it. It must be admitted, however, that this is better than not pressing at all, but it
can only be excused on the plea of enforced economy,
or want of appliances.

The GILL HOT-ROLLING MACHINE is now frequently to be found in the warehouse. It consists of
two large bright cylinders fixed parallel in a substantial
side-frame. Live steam is admitted into each cylinder

at intervals during work, by which means they may be heated to any desired temperature. The laying-on board is placed immediately behind the back cylinder,

Fig. 121.—The Gill Hot-rolling Machine.

and the single sheets are fed in, passing between the rolls, and carried by a series of endless tapes to the taking-off board in front. The pressure is regulated by powerful set-screws under the laying-on board. Theoretically, in the passing between the heated rolls the ink is dried, the marks of impression are taken out, and a gloss is imparted to the paper. It will be readily seen that the rolls must take off a proportion of the wet ink, but to prevent the soiling of the succeeding sheets they are cleaned by means of a pad, filled with sponge, which presses firmly against the surface on the under-side. This pad lies partially in a trough or duct, into which is put a quantity of ley, which assists to absorb the deposited ink.

It is necessary to clean out the duct at least every day, as the sponge takes a quantity of ink, and if clean ley is not supplied will of course re-deposit it upon the rolls.

The speed must necessarily depend upon the size of the sheets being rolled, but it may be safely said to be able to turn out 1,000 sheets of double-demy per hour.

In the management of the Gill, care must be exercised in putting on the pressure. If the screw is used too freely, the sheets will be materially thinned, besides which, a higher gloss is imparted than is generally required. The difference between pressing and rolling is, that in the former case the sheet is rendered perfectly smooth without suffering in bulk, while in the latter the texture is frequently affected. This may be readily understood when we consider that in the hydraulic the whole of the surface is under pressure, while in rolling, the immediate point of contact with the rolls is but a fraction of an inch. Therefore the greatest care must be taken in regulating the impression, otherwise a paper of 60 lbs. weight may only handle as if it were but 40 lbs. after leaving the rolls.

This machine is undoubtedly a valuable adjunct to a jobbing office, especially when work is required to be finished in a very short space of time. But it can be hardly recommended for really first-class cut-work, as the body of ink on the cuts is insufficiently dried by the quick contact and pressure, and the applied heat is apt to impart a brown tinge to the blocks.

Hot Pressing.—A process now falling into disuse. One of the reasons for this is, that the paper may be supplied by the maker with such a splendid surface, that any subsequent process for imparting brilliancy is unnecessary. Moreover, the ink used for superfine cut-work is so bright, that pressing by the ordinary means is quite sufficient to impart that appearance to work that used to be peculiar to sheets which had been through the hot plates. If it is desired to give an extra gloss (a quality held by some to be objectionable), rolling between zinc plates is really all that is necessary. This, in fact, is in many respects better than hot rolling, as the latter is apt to turn the colour of the ink, unless great care is exercised in the heating of the plates.

The process consists in placing thick hot iron plates at intervals between the work which has been fed between glazeboards in the ordinary way, and pressed in the hydraulic.

As an economical substitute, the Gill rolling machine may be used.

The construction of CUTTING MACHINES has of late been considerably improved. The original "Guillotine" is slow in its action, besides requiring considerably more manual labour in the working. The paper is secured by a clamp attached to a large horizontal wheel, with handles at intervals, and turned up or down by hand. This of course is a long process, especially with jobbing work, requiring careful and frequent adjustment. The action of the knife is also comparatively slow.

With the SELF-CLAMP machines the cut is accomplished

FIG. 122.—HARRILD'S SELF-CLAMP CUTTING MACHINE.

in less than half the time, and the clamp, or rather an independent flat bar forming the under-side, may be moved down on to the paper by the action of the foot pressing upon a treadle under the table, or by a convenient hand-lever. The immediate spot upon which the knife will fall may thus be indicated in a moment. When the foot releases the treadle, the bar returns to its position under the clamp—a counter-weight exercising the necessary leverage. As the strap is struck-on for cutting, the clamp and knife act together, the former, of course, securing the work immediately before the descent of the knife. The cut occupies about six seconds. With the aid of this machine the workman can cut at least fifty per cent. more paper, with considerably less labour, than on the original Guillotine, the Self-clamp being practically automatic. In order to force the machine into immediate full speed, a friction clamp is employed, being pressed against the travelling fly-wheel by powerful springs.

Self-clamp machines are now supplied by all the principal printers' engineers, those of Messrs. Harrild, Messrs. Dawson, Messrs. Salmon, and Messrs. Furnival being perhaps the best known. In construction, though varying in minor details, the principle may be said to be the same. Advantages are claimed in each case, either on account of speed, solidity of construction, or economy, but they are all well tried and excellent machines.

Small cutting machines are frequently turned by hand. Although, in some cases, we are aware that the application of power is inconvenient, hand-labour cannot be commended, either on account of speed or of the success attending the operation. If the pile of paper is very thick, the great resistance to the knife renders its passage uneven, and the cutting is not so clean. Cards are difficult to cut by hand, especially when varnished. In this case only a small lot should be attempted each time.

Cutting-sticks, which fit into the table immediately under the knife, are generally made of hard wood, and

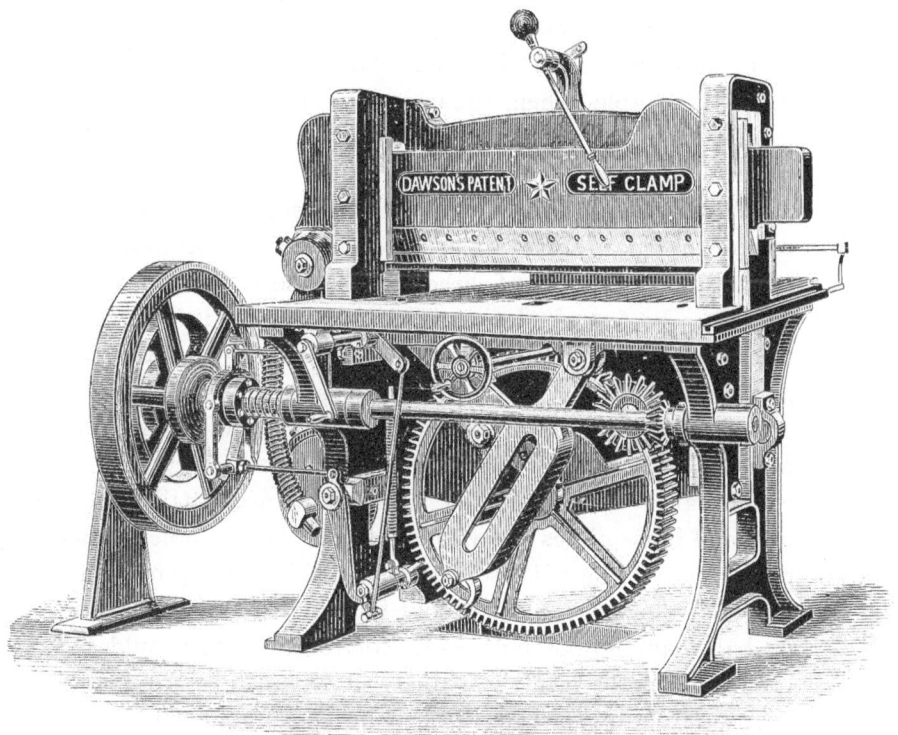

Fig. 123.—Dawson's Self-clamp Cutting Machine.

are supplied by the machine-maker. It may, perhaps, be stated that stereo metal sticks are sometimes used, and may be considered more economical, inasmuch as after they are worn out they may be recast, the actual waste of material being very little. All that is necessary is to have an iron mould made for casting pur- poses. When carefully used they last for some considerable time. If properly adjusted, the edge of the knife is not injured by the contact with the metal.

In the cutting of work, success, of course, principally depends upon the careful knocking-up of the sheets. With thin paper this is a more difficult matter than with stiff stuff, the edge of which falls by its own weight upon the table when squaring the back lay. In the former case, therefore, fewer sheets should be taken at a time.

When sheets have to be cut up into 8vo, especially when worked in quadruple, they should be pointed, excepting in the case of litho work, where the lay is of necessity perfect. The expense of pointing is comparatively small, when we consider that a boy can easily point a ream in an hour. A ream of 8vos in quadruple gives 16,000 copies, so, roughly speaking, the cost of pointing is practically less than a farthing per 1,000, for which an equal margin can be ensured. The points are usually made of small brass plates two inches square, with a hole in the centre, into which is soldered a sharp steel pin. The length of the pin should be about four inches. When a ream or half-ream is pointed ready for cutting, it should be removed to the table of the cutting-machine, and the centre carefully marked. Push one end under the knife, and having ascertained, by bringing down the clamp, that the knife will cut exactly at the point, secure the clamp firmly on the work, and carefully remove the points. Make a cut, when the cut edge will be perfectly even, and therefore by knocking-up render the successive cuts absolutely accurate. The number of sheets pointed for a single cut must of course be regulated by the thickness of the paper or card.

PUNCHING AND EYELETTING MACHINES.—In general jobbing offices these simple machines are a necessity for piercing show-cards, "tear-offs" of calendars, for tapes, &c. Tapes, or fine silk cord, are generally preferred to the brass suspenders, the latter being apt to impart a very stiff appearance, besides proving more awkward to hang. Frequently cards, &c., are simply punched, the brass eyelet being dispensed with. When, however, it is necessary to insert the eyelet, the hole has first to be made, and the eyelet placed in position, and clamped afterwards.

The Punching Machine consists of a strong upright bracket fixed to a substantial iron stand The top of the bracket supports an arm, one end of which is attached to a stout bolt working downwards tightly in a bracket extending

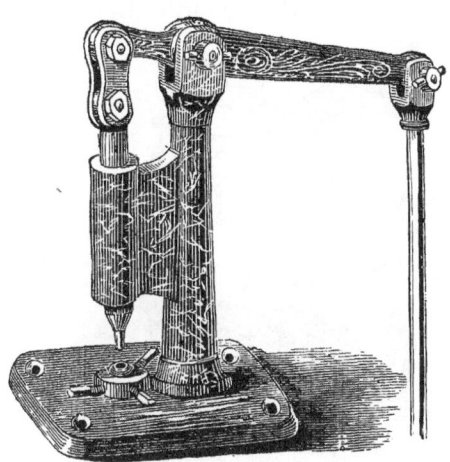

Fig. 124.—Punching Machine.

from the front of the main pillar, and the other to an iron rod running down through the counter to a treadle on the floor. The punch is fastened to the bottom of the bolt. By sharp and sudden pressure on the treadle, cards, strawboards, or millboards are easily and cleanly pierced, great power being obtained from the long lever-arm at top. At the back and on either side of the punch are movable gauges, that the exact spot on the card may be accurately and quickly pierced. Self-feeding eyeletting machines are usually adopted.

The construction of these machines varies slightly from

C C

that of the one just mentioned.   In this case the lever is at
the base of the machine and the bolt is forced *upwards.*
At the end of the movable bolt is a small nipple, working
on a spring.   The eyelets are supplied from a small
circular brass box at the back, and travel down a
narrow groove to a point immediately above the nipple.
The hole in the card is placed on the nipple, and by
means of the treadle the latter is forced smartly up-

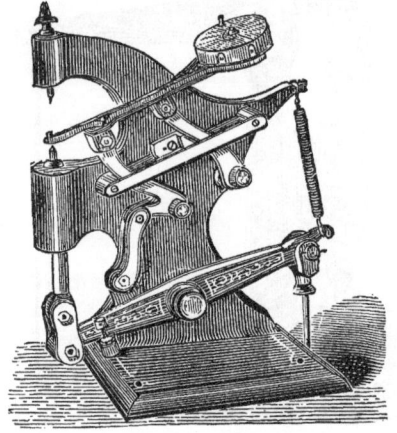

wards and comes into con-
tact with a fixed punch
which forms a burr. When
it is desirable to put eyelets beyond two inches
from the edge, a special
machine is necessary, as
the supporting bracket at
the back must be a greater
distance from the nipple,
to admit of the card being
pushed farther in.   In this
case the eyelets are usually
placed on the nipple by

Fig. 125.—Eyeletting Machine.

hand, which takes, of course, much greater time than
on the automatic machine.

VARNISHING AND GUMMING MACHINES.— Unless
there is a large and continuous supply of varnishing, it is
by far the best plan to give it out to those who make this
class of work a specialty.   Although the machine itself
occupies but little space, a large room fitted with shelves
for the storage of racks is required for the drying process, as each newly varnished sheet must be placed in
such a position as to be entirely free from contact.
Cold air is to be preferred, for if the temperature is too
high, although the varnish may harden on the surface, it

has a tendency to retain its soft body, and does not eventually possess that enamelled brilliancy which should be characteristic of well-finished work. It may also be mentioned that fire risk is materially increased by the storage of open racks of highly inflammable sheets lying about, and this should be a strong argument in favour of obtaining assistance with this class of work.

BROWN AND RAWCLIFFE'S VARNISHING MACHINE in appearance is somewhat similar to the old Desideratum machine, consisting of one large cylinder, upon which the sheet is fed-in. The gripper (which extends in one piece across the cylinder) takes the sheet, and when at the bottom it is brought into immediate contact with a roller, the under-side of which is in a long narrow trough, somewhat similar to the ductor. As the sheet passes between the large cylinder and the smaller roller, a regulated amount of varnish is deposited, and the work is released immediately upon the taking-off board, and laid singly between racks and allowed to dry. The ductor or trough is supplied with varnish from a large tank in the front of the machine, and the outflow can be conveniently regulated.

Messrs. Newsum, Wood, and Dyson also manufacture an admirable varnishing machine. It is self-delivering, and will finish a quarto or quadruple sheet with equal success.

Messrs. Greaves, of Leeds, have also several machines at present in use.

Varnishing machines may be used also for gumming. It is necessary, however, that the whole be thoroughly cleaned before substituting the gum for varnish.

Gumming and varnishing by hand are both dirty and tedious processes, at the same time offering no advantage in appearance as compared with machine work.

C C 2

PERFORATING MACHINES are of necessity largely used in commercial houses. They are of very simple construction—consisting of a table with a slight head-piece extending from side to side, and in which the steel punches are fixed. The head-piece is attached to steel rods on either side, and is forced down by a treadle working a pair of cams. Immediately beneath the punches is a bed-plate, with small perforations, which admit of a free passage of the punches, at the same time keeping them rigidly straight, and preventing their bending. Movable side and back lays are provided on the table.

PACKING.—It must be admitted that frequently too little attention is given to the proper packing of work. This is a mistake. Nothing speaks more eloquently as to the manner in which an office is managed than a newly delivered parcel. When torn wrapper-paper, from motives of economy, has been used for packing purposes, it is liable to create a suspicion in the minds of the recipients that the contents have been executed in a slovenly manner. Parcels of paper are liable at times to be roughly used, in consequence of their weight, and therefore every precaution should be taken that each parcel is firmly and securely tied up. The work should be first well encased in clean paper, and the whole neatly placed in clean brown, and well secured by strong cord. A well-written label should be affixed at one end, stating name of work and number of sheets. If it be necessary to send also the spoilage, it should be made up into a separate parcel, and so labelled. If defective sheets are " mixed in " with the bulk, the printer is liable to be perpetually suspected—a very undesirable result.

CHAPTER XXVII.

ON ERECTING AND DRIVING MACHINES.

General Matter with Reference to Pits, Shafting, &c.—Amount of
Power calculated to drive various Machines—Straps, Riggers,
Counter-shafts, &c.

IT is always advisable that a printing machine be placed
on the ground-floor or basement of a building, where there
is least probability of vibration. With perfecting machines
this of course is a necessity, as pits have to be provided;
but in the case of a Wharfedale, where possible, it is a good
plan to have a pit, as the working parts may be more conveniently lubricated, watched, and generally attended to.

The average depth of a pit is 2 ft. 6 in. It is
usually lined with bricks, and the edge upon which the
side and top and bottom frames are placed should be
stone, about 12 in. wide, sunk level with the floor.
The bottom is concreted to keep out the surface water.
Sometimes the pit is made shallower than the dimensions
given above, which may be rendered necessary by the
dampness of the soil.

When machinery is of necessity placed upon wooden
floors, unless specially supported by iron girders, the
vibration will always be marked. Frequently space is
utilised which is ill adapted for the purpose, the result
being that the machines as well as the work suffer. It
is patent that a machine will always work more satisfactorily upon a solid foundation than upon a simple
floor. The shaking will not only affect the register, but
is often liable to materially increase the cost of necessary
repairs. Wherever there is a perpetual jar there will be

unequal wear, and when this is the case really satisfactory results are out of the question. For ordinary
commercial work, this perhaps matters little, so far as
the work itself is concerned, but machines are frequently
condemned as not being well made and balanced, when
the actual fault is entirely caused by the conditions under
which they are worked.

When it is found necessary to erect a machine upon
a floor, an expert should be consulted as to the desirability, not only of strengthening it, but of securing the
beams in such a manner as to render it as rigid as
possible. Machines are heavy in themselves, and many
tons in dead weight occupy but a small floor-space ;
but when the machine is put in motion the strain is
materially increased, and unless in the first instance such
precautions are taken, the vibration becomes very great.
In laying out the plan of a machine room, the machines
should be placed as nearly centrally over the girders as
possible ; and it is even then desirable to "tie" the
girders up by means of stout iron rods—extending from
one to the other—which will to a great extent prevent
excessive vibration.

Of course, we are aware that it is impossible in all
cases to adopt the best system, as rooms are frequently
built for any purpose rather than for the erection of
machinery. But in drawing out a plan, we have not
only to economise space, but to take into consideration
the most convenient way in which the driving power may
be transmitted, and to obtain the best light. It depends
entirely upon the shape of the room as to the position
the machines should occupy, and consequently how the
shafting may be hung.

Wharfedales, and in fact all machines which do not
of necessity require a pit, are generally placed upon

substantial wood planks. This not only prevents the vibration from the other machines being felt so much, but insures a more solid and level foundation, and to a great extent diminishes the effect of its own movement on the building.

SHAFTING in every case should be well supported. It is very rarely possible to carry it from one extremity of a room to the other without intermediate bearings, and as a matter of necessity hangers are placed at short intervals. There is a great "lug" with large machines, and if the shaft is insufficiently supported it will certainly give ; and nothing we know of is more irritating than to see a length of shafting rotating out of its centre.

Periodical examination should be made of the bearings, that the lubrication may be perfect. The needle lubricator is generally adopted. This provides a constant supply of oil to the journal when the shaft is in motion. Capillary attraction stops the flow, however, when the shafting is not running.

Fig. 126.—
Needle
Lubricator.

The fixing of the shafting must necessarily depend upon circumstances. We are strongly in favour of having it under ground, but this is not convenient unless excavations are specially made when the building is being planned. Besides being less dangerous generally, this system materially adds to the neatness of the machine room.

When machines are driven on the floor above, in addition to the basement, of course the shafting should be so arranged as to drive both the rooms, the straps being carried through the flooring. The only objection to this is, that the straps running through the floors are of necessity apt to be full short—which is a continual source of trouble, as, in addition to the limited travel (especially in the case of large riggers), they are apt to frequently break, in consequence of their being compelled to be laced more tightly than those of a longer run.

Again, tight straps are a decided danger to the machine. If by any chance a cog breaks and jams, the unyielding band will possibly cause a very serious break; whereas, if it were moderately loose, the sudden jar would throw it off the pulley and prevent further mischief.

Various methods are adopted to prevent the oil dropping from the bearings. It is advisable to fix under every hanger a small tin tray to catch the oil that may fall. Sometimes these trays are constructed with a " dip " in the centre, with a short pipe and cock fitted at the bottom. The waste oil that has accumulated can then be drawn off by opening the tap. It is needless to state, perhaps, that however much may be collected, it should be thrown away, as it is sure to be full of grit and dirt.

Shafting should be frequently cleaned, which may be safely and effectually done by means of a long wooden pole, having at the end a hook with a broad flange. Fasten a piece of oiled cotton waste on the inside round the hook, and hang it over the revolving shaft, exerting pressure by pulling at the other end. The shaft, in close proximity to the riggers, and the inside of the riggers, should be wiped out when the shafting is stationary.

When power has to be transmitted to a distance, say from the bottom of a building to the top, bevel wheels are frequently employed. In this case the teeth of one wheel is made of hornbeam, in order to deaden the sound and prevent extensive vibration. Sometimes this method of driving shafting is unavoidable, but when possible it is by far the better plan to employ two, or even three, intermediate shafts rather than adopt the spur wheels. The strain at the points where the cogs are in contact is excessive, and a large amount of power is lost. In addition to this, the teeth will sometimes "top," owing to a sudden strain, and the wooden cogs are of course destroyed. This is a source of continual danger and expense. The dull and perpetual noise is also objectionable, and as the teeth wear, as they assuredly will, and the play becomes great, this evil increases.

It is somewhat difficult to state the exact amount of power required to drive the various machines. Without doubt the ordinary platen requires more than any of the other book-work machines, owing to the strain exerted on the impression. It will be noticed that, when the platen is being pulled over, the strap will labour considerably, and the shafting receives a check. In fact, it may be safely stated that a sudden strain is exerted equal to between three and four horse-power when the impression is being given on, say, a double-demy platen. Of course this occurs only at the moment when the platen is on the forme, and therefore the strain is spasmodic, not continuous, but proves, nevertheless, very trying to the engine. In consequence of this, platens should never be driven off the main shaft, but from a special counter-shaft. Rolling machines should also be driven from an independent shaft, as when the plates enter the rolls the sudden jar is apt to momentarily pull up the shafting, in which case the register will be affected very materially on all perfecting machines driven off the same length.

These remarks do not apply to the Napier platen, the power required for the impression, in consequence of the knuckle-joint motion, being comparatively small.

The difficulty of calculating the power necessary to

drive a machine of a given size must necessarily depend
upon the dimensions and style of forme being worked.
For instance, if a light tabular forme is being printed,
the machine requires just sufficient power to turn it
round ; but if a heavy forme of cuts, or a large, flat,
and solid tint is being worked, the force required to pull
the impression, without check, will be at least ten times
as great.

Generally, with an average forme, it may be stated
that to work a

Double-demy Wharfedale, about $\frac{1}{2}$ h.p. is required.
4-Demy ditto   „   1 h.p.   „
4-Demy ordinary Perfecting „   2 h.p.   „
4-Demy Anglo-French   „   2 h.p.   „
Double-demy Platen   „   3 h.p.   „
Double-demy Napier ditto   „ $\frac{1}{2}$ h.p.   „

In calculating the total amount of power required to
drive a given number of machines, although it is advisable to have a good surplus, it should be remembered
that it is a very rare occurrence to have the whole
running at the same time.

STRAPS.—Those usually adopted are 3-inch single.   In
the case of large machines 4-inch are required.   Ordinary
platens should be provided with 4-inch double, otherwise frequent breakages will occur.   It is advisable that
they be kept perfectly tight, or the machine is apt to
pull up on the impression.

A new strap will require to be re-laced several times
before all the stretch is taken out.   If it is liable to slip,
dust a little powdered resin on the inside.   Ink is sometimes used, but this is a clumsy and dirty expedient.

A strap should always be narrower than the rigger.
It must not be allowed to overlap the edge, as in the

first place it will cut through at the end of the flange, added to which, power is lost, as a portion of the surface of the strap is doing no work. With cone pulleys, if the strap is full wide, one side works against the side on one extremity, and rubs and overhangs on the other.

The width of engine straps must in all cases depend upon the power of the engine and the width of the driving pulley. They are generally riveted with flanged gun-metal pins. For tightening them the screw apparatus is found by far the most effective. It consists of two sets of parallel bars working on a long screw. When this tool is used the strap may be placed in position over its riggers, and the two ends to be joined nipped firmly between the parallel bars. With a stout "tommy" the long screw may be turned, forcing the ends of the strap closer together, and when sufficiently taut, the rivets are fixed through the holes made by a punch.

Rolling machines require substantial straps. Whenever possible, a long travel should be allowed, as the jar when the plates are entering the roll does not then affect the shafting to such an extent as when the strap is comparatively short. If the band is short, it must necessarily be rather tight, or it will slip round the riggers. If it is too taut, however, the shafting will sustain a check when the plates enter, which in the case of a longer strap is to a great extent counteracted by the swing. Countershafts should be employed, as in the case of the platen machine, where possible, that the strain may to some extent be at least divided.

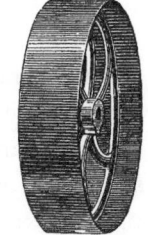

Fig. 127.—
Rigger or
Pulley.

COUNTERSHAFTS are short pieces of shafting supported by hangers driven by a band from the main shaft. Sometimes they are employed in consequence

of the situation of the machine to be driven being such as
to render the travel of a strap from the main shaft inconvenient.   When it is found desirable to drive a machine
at different speeds, to suit varied work, a countershaft
is, of course, indispensable.   On the main shaft is fixed

Fig. 128.—Countershaft with Three-Cone Pulleys.

a cone, generally of three different diameters, and a similar
cone on the countershaft, the ends, however, being reversed,
i.e., the largest flange facing the smallest on the main
shaft.   It will thus be understood that the strap, once
properly adjusted, is exactly the length necessary for
either of the three speeds, for if it is moved to a smaller
rigger on the main shaft, it is shifted to a larger on the
countershaft.   These cones are generally calculated to run
the machine 400, 750, and 1,000 per hour.   The speed
must, of course, be decided by the class of work being

printed. In houses where the work is of a very miscellaneous nature, it is necessary to frequently alter the
speed of the machine. A ladder should be specially
provided for this purpose, having iron hooks fixed at
the top of each side to fit loosely over the shaft. By

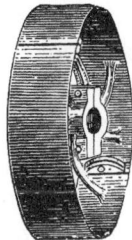

Fig. 129.—Split Pulley.   Fig. 130.—Spring Split Pulley.

first pulling the strap from off the cone pulley on the
rigger, the countershaft is of course stopped. The strap
may then be readjusted on the proper rigger of the

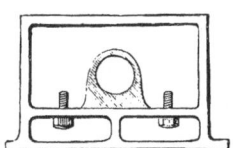

Fig. 131.—Wall Boxes for supporting Extremities of Shafting.

countershaft, and run round the corresponding rigger of
the main shafting. Sometimes, however, a long piece
of iron piping is employed to remove and readjust the
strap.

With reference to riggers or pulleys, split pulleys, or
those made in two pieces and bolted together after
adjustment, are to be preferred, as they may be easily
removed, or shifted to a different position with little
trouble. It is easy enough to run solid pulleys upon

the shaft prior to the latter being fixed in position, but a very different matter when the shafting is fast in its bearings.

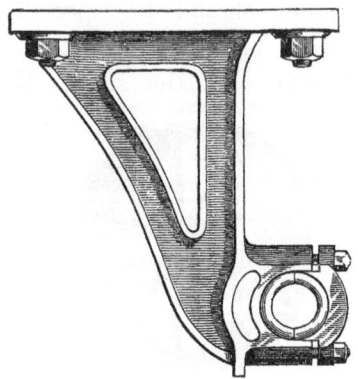

Fig. 132.—Hanger.

When it is necessary to employ wide-flanged pulleys, say, upwards of nine inches in width, either a double set of spokes should be provided, or two narrow riggers placed close together may be employed, as the strain generally is on *either side of the spokes.* We have known instances in which the riggers have been split or crushed from want of proper support at the extremities.

Where shafting is subjected to great strain, square brackets should be used as a support instead of the ordinary hanger, as the latter are liable to give actual support only on one side. This is especially necessary when bevel wheels are employed, to drive another shaft at

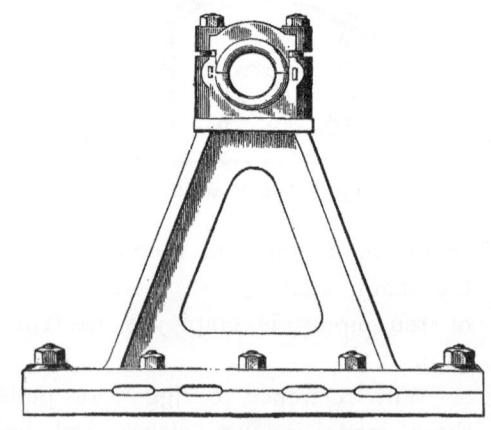

Fig. 133.—Pedestal.

an angle, otherwise the great lug will inevitably strain the shafting.

Hangers or brackets should be firmly fixed to a sound block of wood, about four inches thick. The block must be at least six inches larger either way outside, to admit of the bolts having plenty of space. The bolts should be carried through the floor or beam, and be well secured by nuts on the top side.

When the shafting is carried beneath the ground-floor, it is supported by pedestals, or inverted hangers.

CHAPTER XXVIII.

POWER.

The Beam and Horizontal Engines—Fuel—The Otto Gas Engine.

IT is not our intention, in this chapter, to describe the construction of the steam engine, the properties of steam, or the treatment and management of boilers, as there are many popular and admirable works now published dealing exhaustively with these subjects.

With reference, however, to the choice of steam engines for driving printing machines, there can be but little doubt that the beam engine is to be preferred. The reciprocating motion of the beam itself adds materially to the general uniformity of speed. With the book printer this is a matter of very considerable importance, as the register of the work is liable to be affected by any irregularity, more especially in the case of Anglo-French machines. With newspaper work, however, the consequences are less serious.

The matter of space very often influences the decision. The beam of course requires more height than a horizontal, and inasmuch as the lightest and best positions are generally occupied by printing machines, the engine is frequently erected in a small and inconvenient place. This, we venture to think, is unwise. If the engine driver has to work in a confined space, and experiences difficulty in attending properly to his engine, he is apt to become careless, and the machine which should always be the brightest and cleanest in the establishment will

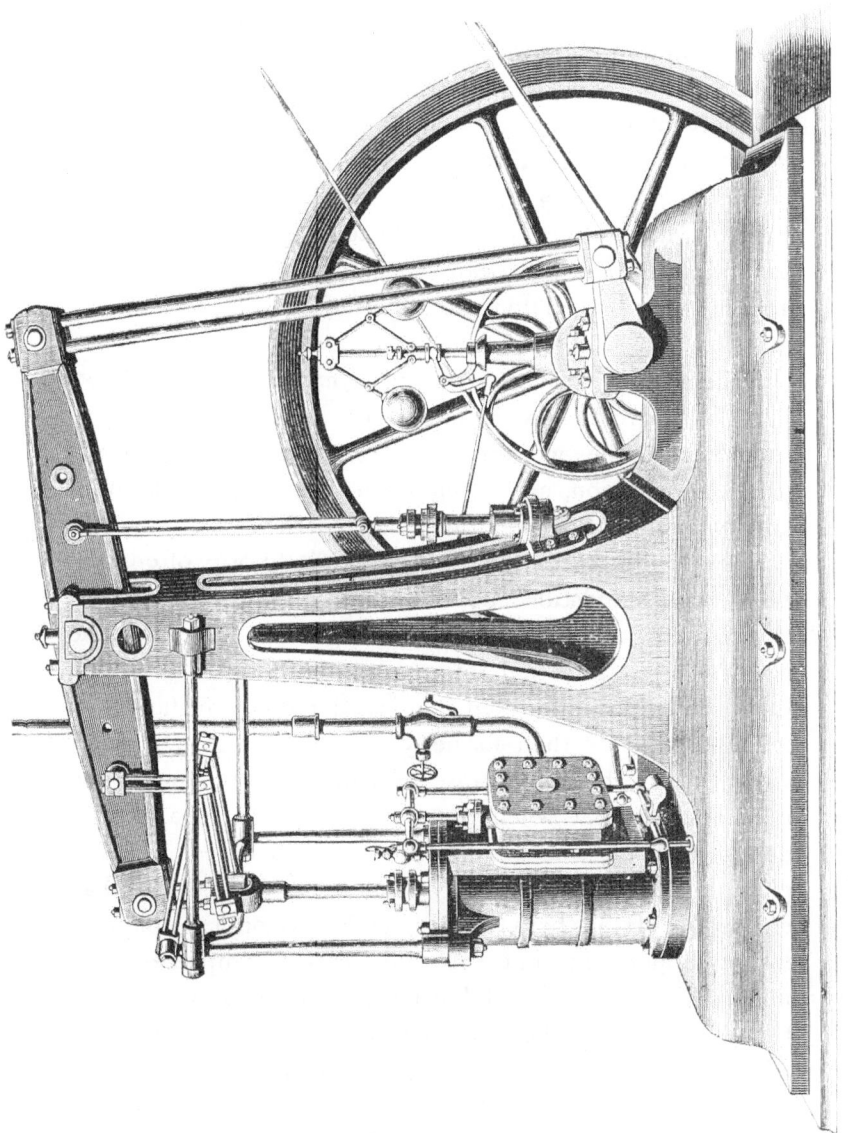

FIG. 134.—BEAM ENGINE.

D D

often present but a sorry appearance. There are few more pleasing sights than a bright and quiet steam engine ; and to the man in charge it should be a matter not only of duty but of pride to have everything about the room clean and tidy.

When comparing the beam and horizontal engines, although the former is decidedly to be preferred, it must be admitted that the beam consumes the most steam— it has been calculated at about a sixth or seventh more.

The steam engine rarely gets out of order—excepting from gross carelessness. The construction is so simple, and the workmanship so excellent, that, with proper lubrication, and attention to the brasses, bearings, and packing, but little trouble is experienced.

The boiler, generally speaking, is more often the cause of a standstill. But the improved method of setting has to a great extent simplified matters ; and the periodical examination by experienced men, to which they should be subjected, has rendered them far more reliable and safe than formerly. Corrosion of the plates was perhaps the chief cause of trouble and danger, and it is obvious that when they were enclosed in a case of damp bricks it was a difficult matter to detect a faulty plate. Now, however, they are rather supported than enclosed. It is both a safe and economical plan to insure boilers in one of the many companies now undertaking this class of business. The inspectors whom they depute to make examinations are, as a rule, to be thoroughly relied upon, and if any repairs are necessary, an official intimation from the company is soon forthcoming.

As regards fuel, there is no question as to Welsh coal being by far the best, both for its heating properties,

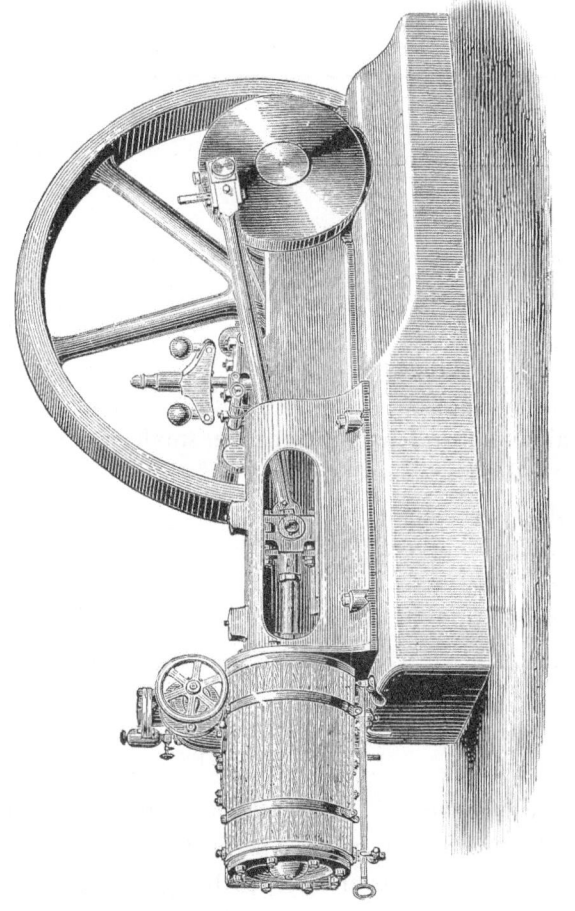

FIG. 135.—HORIZONTAL ENGINE.

D D 2

and because, with proper stoking, it is almost smoke-
less. The following table, compiled by Mr. Wicksted,
giving the comparative ratios of evaporation, may be of
interest :—

Name of Fuel.	Water evaporated from 52° by 1 lb. of fuel.
	lbs.
Welsh, Best	9·493
Anthracite	9·014
Newcastle, Best small	8·524
„ Average	8·074
Welsh, Average	8·045
Gas coke	7·908
Half coke and half Newcastle small ...	7·897
„ Welsh „ „ „ ...	7·865
„ Newcastle „ Derbyshire ...	7·710
Newcastle, Average large	7·658
Derbyshire	6·772
Blythe Main Northumberland	6·600

GAS ENGINES.

It is frequently urged that steam is on the whole
more reliable than gas. The justice of this can hardly
be sustained, however, when the gas engine of recent
construction is taken into consideration.

Gas engines are being very largely adopted of late,
and with the most satisfactory results. Generally speaking, they may be considered far more economical in the
working than the steam engine. They occupy but little
space. while constant attention on the part of a man is

unnecessary. Not that the engine should be handed over to the mercies of some irresponsible individual. This plan frequently results in the stoppages often complained of. One man should have whole and sole charge, and although employed upon other work, he should always be near his engine. The gas engine is a more delicate piece of machinery than the steam engine, and requires careful handling. It must be kept thoroughly clean and well lubricated. The amount of oil required may appear excessive, but one of the secrets of the gas engine running successfully is the plentiful supply of oil of a *proper quality*. This cannot be too strongly impressed. A special oil is recommended by the makers, of which we would advise the use.

The OTTO gas engines may without doubt be considered the most successful. They are made in all sizes, from two-man-power to forty-horse-power nominal. With reference to the cost of driving, it is a decided point in their favour that expense is only incurred while they are actually at work. There is no risk from fire, and this fact is recognised by the fire insurance companies, for no extra premium is imposed wherever they may be erected. The driver has merely to look to the lubrication, light the jet, and assist the fly-wheel, when the engine starts at full speed. In the case of the steam engine the stoker has to spend some time in getting up steam, and also in drawing the fires at the end of the day; added to this, consumption of fuel must go on during meal times. It has been calculated that, with gas of good quality, the consumption is about eight cubic feet per hour per indicated horse-power.

Notwithstanding that the makers of the Otto profess that it is a " silent " engine, it may be questioned if this adjective can, strictly speaking, be applied, although, in contrast with some of its predecessors, there is no doubt that it is comparatively quiet in its working. The noise caused by the escape of the exhaust gases is greatly minimised by the adoption of the "quieter" which Messrs. Crossley fix to the end of the pipes.

Fig. 136.—The Otto Gas Engine.

In all cases it is advisable to have wrought-iron pipes fitted for carrying off the exhaust, as it sometimes happens that a charge of gas will escape unexploded. In its passage to the open, this gas may explode in the super-heated pipes, which, if not thoroughly reliable, will become fractured. We mention this, as we have known accidents to occur from this cause.

In conclusion, it may be stated that the Otto is well adapted for the driving of printing machinery, not-withstanding the fact that the impetus is only given at one end of the cylinder. But the speed at which they are driven (about 160 revolutions per minute), assisted by substantial fly-wheel or wheels, insures a steady motion, which overcomes a varying demand, and the governors are so nicely adjusted that little variation is detected.

It is true that the first cost of the Otto is greater than that of a steam engine of the same power; but the former requires no boiler, and no structural alterations are necessary for its accommodation; and when we come to include the boiler with the steam engine, and further add to the account not only the original cost of purchase, but also the expenses of driving and of repairs, we are inclined to think that, so far as economy is concerned, gas may claim a general advantage over steam for driving purposes.

THE END.

Printed by Cassell & Company, Limited, La Belle Sauvage, London, E.C.